THE SINGAPORE
RESEARCH STORY

World Scientific Series on Singapore's 50 Years of Nation-Building

The complete list of titles in the series can be found at
http://www.worldscientific.com/series/wss50ynb

World Scientific Series on
Singapore's 50 Years of Nation-Building

THE SINGAPORE RESEARCH STORY

Editors

Hang Chang Chieh
Institute for Engineering Leadership, NUS, Singapore

Low Teck Seng
National Research Foundation, Singapore

Raj Thampuran
A*STAR, Singapore

World Scientific

NEW JERSEY · LONDON · SINGAPORE · BEIJING · SHANGHAI · HONG KONG · TAIPEI · CHENNAI · TOKYO

Published by

World Scientific Publishing Co. Pte. Ltd.

5 Toh Tuck Link, Singapore 596224

USA office: 27 Warren Street, Suite 401-402, Hackensack, NJ 07601

UK office: 57 Shelton Street, Covent Garden, London WC2H 9HE

Library of Congress Cataloging-in-Publication Data
Names: Hang, Chang C., editor. | Teck, Seng Low, editor. | Thampuran, Raj, editor.
Title: The Singapore research story / editors, Hang Chang Chieh,
    Institute for Engineering Leadership, NUS, Singapore, Low Teck Seng,
    National Research Foundation, Singapore, Raj Thampuran, A*STAR, Singapore.
Description: 1st Edition. | New Jersey : World Scientific, 2015. |
    Series: World scientific series on Singapore's 50 years of nation-building |
    Includes bibliographical references and index.
Identifiers: LCCN 2015038918 | ISBN 9789814641258 (hardcover : alk. paper)
Subjects: LCSH: Singapore--Economic policy--21st century. | Singapore--Economic conditions--
    21st century. | Technological innovations--Economic aspects.
Classification: LCC HC445.8 .S5718 2015 | DDC 330.95957--dc23
LC record available at http://lccn.loc.gov/2015038918

British Library Cataloguing-in-Publication Data
A catalogue record for this book is available from the British Library.

Typeset by Stallion Press
Email: enquiries@stallionpress.com

Printed in Singapore

Foreword

In 1985, the Singapore economy was in recession. For the first time, our domestic economy contracted while the global economy was still growing. We established that this was in part due to high wage costs which had eroded our competitiveness. The recession was thus not only cyclical, but revealed more basic problems within our economy.

We convened the Economic Review Committee to review our policies and strategies. In February 1986, we published a report "The Singapore Economy: New Directions" to explain the causes and chart a new course. One key strategy was to invest in Research and Development (R&D) so as to build a more resilient, knowledge-based economy.

The National Science & Technology Board (and subsequently its successor the Agency for Science, Technology and Research), together with the Economic Development Board and other agencies were tasked to implement the R&D strategy. Our focus on R&D helped Singapore to become the preferred partner for leading MNCs and universities to establish R&D centres and other high value-added activities, creating good jobs in both research and other complementary roles. It has also benefitted Singaporean enterprises and SMEs, by encouraging companies like Keppel Offshore & Marine, Advanpack Solutions Private Limited and Singapore Asahi Chemicals and Solders Industries Pte Ltd to create innovative products/services.

R&D has also helped us to overcome our constraints and meet national needs. Water is one success story. PUB's R&D programme has given Singapore two new sources of water — NEWater and desalinated water, complementing the traditional sources of water — imports and local catchments. The capabilities we have built in water research have helped grow the water industry. Today, we have a thriving water ecosystem with over 180 water companies and 26 research centres. These include not only Singapore-owned companies such as Hyflux and Sembcorp, but international names like Black & Veatch and Meidensha.

Our universities have become world-class research institutions, especially after the formation of the National Research Foundation in 2006. We have encouraged creativity and innovation in our young people, and opened up career opportunities in Science and Technology. We now have a pool of highly talented and skilled people — a good mix of local and foreign talent — in Science and Technology and a vibrant entrepreneurial community.

I would like to thank all those who have played a role in the Singapore Research Story. In particular, I would like to thank Dr Sydney Brenner who helped to establish the Institute of Molecular and Cell Biology (IMCB) in 1985, and Professor Hang Chang Chieh, who made key contributions towards establishing key R&D infrastructure in Singapore.

I hope that in the next 50 years of our Research, Innovation and Enterprise journey, we will scale new heights in research excellence and create more value in Singapore. As the Singapore Research Story shows, the spirit of R&D is that of can-do and infinite possibilities. Anything is possible, if only we persevere and put our minds to it. I hope that spirit will drive not just our R&D folk, but all Singaporeans as we write the next chapter of our Singapore Story together.

Congratulations on telling this inspiring Story.

LEE Hsien Loong
Prime Minister of Singapore

Contents

Acknowledgements

This book is the product of a multi-agency approach — like our R&D ecosystem today. It could only be written with the help of many, only some of whom are named or the list will be too long. Some 30 people from NSTB/A*STAR and its research institutes were interviewed for their insider knowledge of how the institutes came about and were developed in the early years. We would like to thank Dr Bill Chen for his thorough write-up of his experience from which we were able to extract a shorter account for this book.

The whole project of writing up the Singapore Research Story came together through the support of A*STAR and its Office of Science Communications & Archives (OSCAR) Director Choong Ket Che whose staff arranged and facilitated the interviews and who found the funding for this project. Hang Chang Chieh was granted special approval to access confidential records which helped him to confirm some historical decisions in NSTB/A*STAR. We would especially like to thank Lee Geok Boi who researched, interviewed the research scientists and engineers, and drafted many parts of the manuscript. Numerous annual reports were ploughed through for the official viewpoints as well as newspaper clippings, Internet sources, and private conversations.

We would like to thank the staff of particularly the Economic Development Board for fact-checking, SERC and BMRC of A*STAR for putting together some of the highlights of their institutes. We would also like to thank Prof Lee Eng Hin, Tan Gee Paw and Choo Chiau Beng, for backfielding parts of the draft manuscript, and A*STAR chairman Lim Chuan Poh for reading through the final manuscript. The views expressed are our own and may not necessarily be the official viewpoints. It is our hope that our readers be they young people thinking of a career in R&D or policy-makers will be able to find lessons in our Singapore Research Story.

Hang Chang Chieh (National University of Singapore),
Low Teck Seng (National Research Foundation)
and Raj Thampuran (A*STAR)

Cover Photo Credits

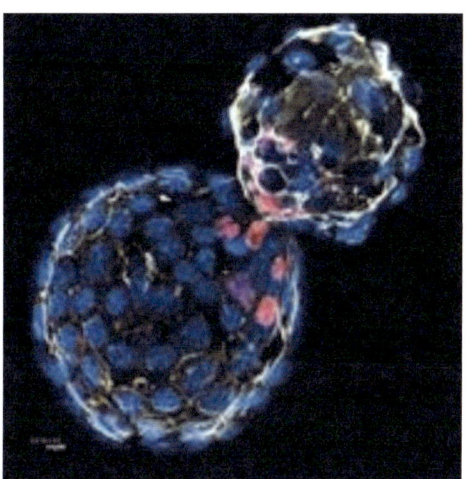

This microscopic image shows a 4.5-day old mouse embryo that has multiplied into 128 cells of three different types. Cultured without a supporting matrix, the usual oval-shaped morphology has been lost as the embryo approaches the implantation stage of its development. Work by Roy Teo, Genome Institute of Singapore, Agency for Science, Technology and Research, in collaboration with the University of Cambridge.

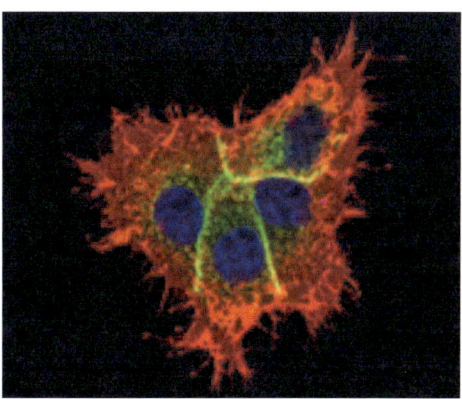

A biomarker which would help physicians to predict how well cancer patients respond to cancer drugs by monitoring PRL-3 and EGFR tumour proteins that are frequently associated with breast and lung cancers in humans. This image shows how cancer cells project multiple extensions on their surface (visualized in red) to gain motility. PRL-3 (visualized in green) enhances this process by inducing activation of EGFR. This research led by Prof Zeng Qi from the Agency for Science, Technology and Research's Institute of Molecular and Cell Biology has given new hope for treating PRL-3 driven cancers successfully with their humanized PRL-3 antibodies for effective cancer therapy.

Graphene is a one-atom-thin crystalline form of carbon in the shape of a honeycomb lattice. Developed by scientists from the National University of Singapore's Graphene Research Centre, this picture shows how hydrogen atoms can be chemically attached to the graphene structure to become magnetic. This was published in *Nature Physics* 9, 284–287 (2013) doi: 10.1038/nphys2576.

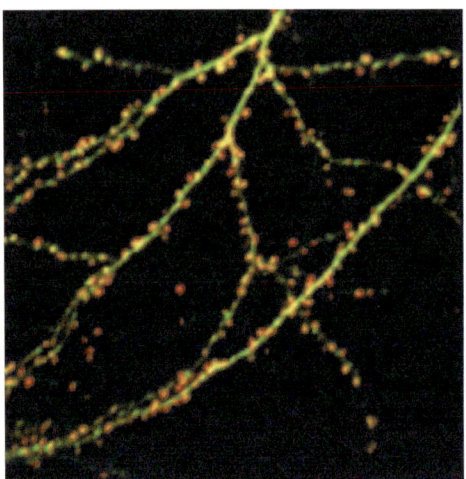

Research is underway by scientists from the Institute of Molecular and Cell Biology, Agency for Science, Technology and Research, to identify the precise role of the protein, SNX27, in the pathway leading to memory and learning impairment. This image shows a brain nerve cell studded with SNX27, which are the yellow-to-orange bulb-like structures, to explain defects in the cognitive development of those with Down's syndrome.

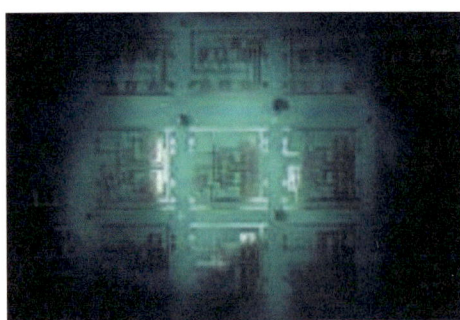

Low-cost printed-electronics on flexible substrates like plastic, paper and cloth are now made possible by researchers from the Nanyang Technological University's School of Electrical and Electronic Engineering. This image shows a printed electronic circuit on plastic film printed using a low-cost tee-shirt printer.

About the Contributors

Bertil Andersson is a plant biochemist from Sweden, and is the author of more than 300 papers covering topics ranging from photosynthesis research and biological membranes to light stress in plants. Prof Andersson was installed as NTU's third President on 1 July 2011 having joined the university as Provost in 2007. He was formerly President of Linköping University, Sweden, from 1999 to 2003, Chief Executive of the European Science Foundation from 2004 to 2007, and Chairman of the Nobel Committee for Chemistry in 1997. He was also Vice-President of the European Research Advisory Board.

Barry Halliwell is the Senior Advisor to NUS President, Tan Chin Tuan Centennial Professor and Leader of the Neurobiology/Ageing Research Programme at the National University of Singapore (NUS). For almost 10 years as Deputy President (Research and Technology), he steered the NUS research agenda and participated in the transformation of NUS into a leading global research-intensive university. An internationally-acclaimed biochemist, he is known for his seminal work on the role of free radicals and antioxidants in biological systems and is one of the world's most highly-cited researchers. He has received numerous research awards, including the President's Science and Technology Medal (2013).

Hang Chang Chieh is Executive Director, Institute for Engineering Leadership and Head of Division of Engineering and Technology Management, National University of Singapore (NUS). He has been with NUS since 1977, serving in various positions including being the Deputy Vice-Chancellor in-charge of research. From 1991 to 1999, he served as Deputy Chairman, National Science & Technology Board. From 2001 to 2003, he was seconded to Agency for Science, Technology and Research to serve as its Executive Deputy Chairman. From 2001 to 2009, he served as Chairman, Intellectual Property Office of Singapore. From 1974 to 1977, he worked as a Systems Technologist in the Shell Eastern Petroleum Company.

Kong Hwai Loong is a Senior Consultant Medical Oncologist in private practice. He was previously the Executive Director of BMRC and Deputy Managing Director for A*STAR. He was instrumental in the establishment of the Biopolis. Dr Kong has won numerous awards, including the National Day Award, two-times winner of the Singapore Youth Award, American Society of Clinical Oncology Merit Award (first Singapore doctor to win this award), and the Singapore Young Scientist Award. He was the past Chairman of the Chapter of Medical Oncologists, Academy of Medicine, and member of the Singapore Youth Award Science & Technology Committee. He is currently a Governor of Raffles Girls Secondary School and Raffles Institution.

Low Teck Seng is a tenured professor at Nanyang Technolgical University and the National University of Singapore. He is a Fellow of the Institute of Electrical and Electronics Engineers and an International Fellow of the Royal Academy of Engineers, UK. He is currently the CEO of the National Research Foundation, Singapore. Prior to his appointment at the NRF he was the Managing Director of the Agency for Science, Technology & Research. He is the founding Principal of Republic Polytechnic and had served as Dean of Engineering at the NUS. He also founded the Data Storage Institute. He was awarded the National Science and Technology Medal and the Public Administration Medal (Gold) in 2004 and 2007 respectively.

Philip Ong was formerly Deputy CEO of the National Research Foundation. He has served in the Public Service Division in the Prime Minister's Office, Ministry of Manpower, Ministry of Education, Ministry of Defence, and Ministry of the Environment and Water Resources.

Png Cheong Boon is the Chief Executive Officer of JTC Corporation. He is a board member of Ascendas-Singbridge Pte Ltd, Surbana Jurong Pte Ltd, Jurong Port Pte Ltd and the Singapore Suzhou Township Development Pte Ltd. He also served as the Deputy Chairman of China-Singapore Suzhou Industrial Park Development Group Co. Ltd and as a member of the management board of the Institute for Engineering Leadership at the National University of Singapore. Prior to joining JTC in 2013, he served as the Chief Executive of SPRING Singapore from 2008 to 2013, and as its Deputy Chief Executive from 2003 to 2008. He started his career in the Economic Development Board in 1993 and held various appointments in Singapore and the US.

Raj Thampuran is the Managing Director of the Agency for Science, Technology and Research (A*STAR) since 2012. He has held various senior leadership positions at A*STAR including the Executive Director of the Institute of High Performance Computing and the Executive Director of the Science and Engineering Research Council. He is a Mechanical Engineer by training and holds a PhD in Materials Science. He also completed an Advanced Management Programme at INSEAD (Fontainebleau). He is an Adjunct Professor of the Nanyang Technological University and an Adjunct Professor in the Dean's Office, Faculty of Engineering at the National University of Singapore.

Tan Kai Hoe is currently the CEO of Accuron Technologies Ltd, a global precision engineering and technology group headquartered in Singapore. Before joining the private sector, he was the Chief Executive of SPRING Singapore, a government agency overseeing Singapore's efforts in enterprise development, entrepreneurship and innovation. He also serves on the boards of several companies and non-profit organisations, including Employment & Employability Institute and Temasek Polytechnic.

Yeoh Keat Chuan is the Managing Director of the Economic Development Board (EDB), the lead agency for investment promotion and industry development in Singapore. He was appointed Managing Director on 1 July 2012. Prior to this, he was EDB's Assistant Managing Director, responsible for investment promotion efforts in a number of industry sectors. He has held several director-level positions in EDB since 2005, overseeing initiatives in divisions including the Americas and the Biomedical Sciences Cluster.

Introduction

Since Singapore's independence in 1965, its leaders have invested tremendous effort and resources to develop the economy in order to create jobs for its people, to grow private sector R&D investments in Singapore and to support nation-building. During the first 25 years, Singapore did not have any urgent need or plans for developing research, although it invested heavily in education and its associated infrastructure in order to scale the technology ladder as the economy evolved from being labour-intensive to skills-intensive and then technology-intensive. Early efforts by public agencies in research and development (R&D) were geared towards helping certain sectors of industry move up the technology ladder and build product development capabilities. The need to compete with the more advanced economies led to a recognition of the need to build indigenous capabilities particularly in engineering and information technologies, and a stronger emphasis on science and engineering, which laid the foundation for R&D. The government's decision in 1990 for Singapore to invest significantly to develop research capability and infrastructure in science and technology ushered in a new era. The next 25 years saw a quantum jump in research investment and accomplishments that soon drew international admiration. It demonstrates that long-term strategic planning, pragmatic policies, forward-looking strategies, political will and execution skills to overcome seemingly poor odds can produce outstanding results in a sustainable way.

This book describes the challenges faced by Singapore in its quest to develop public research institutes, train research scientists and engineers (RSEs) and actively attract collaboration, through public-private partnerships, with companies for industry growth and economic development. The formative steps taken in the first 25 years were against a backdrop of limitations: a weak local talent base, a feeble research culture, inadequate investment in time, money and planning to develop a technology-intensive economy. In 1981, NUS set up the Institute of Systems Science as a teaching institute; in 1985, plans were drawn up for the Institute of Molecular and Cell Biology as a research institute focused on basic science. Then came a major recession in 1985–86 after which an ambitious plan for economic restructuring towards a knowledge-based economy with strong R&D capabilities was outlined in the 1986 Economic Committee Report. A new statutory board, the National Science and Technology Board (NSTB), was created in 1991 and given the mandate to achieve the mission of developing competitive R&D capabilities in sectors important to Singapore manufacturing, and building infrastructure to attract talent and undertake advanced R&D, in close consultation with the Economic Development Board (EDB). Key local university leaders and experienced R&D scientists & engineers from all over the world were recruited and they pioneered the establishment and development of research institutes in engineering and the biomedical sciences. The stellar work of the founding directors and the rapid recognition of the quality of these mission-oriented research institutes gave NSTB the confidence to pursue its mission with more investments. Over the next 15 years, three multi-billion five-year national science and technology plans were conceived and successfully implemented to position Singapore into a research-intensive economy.

In 2002, NSTB became the Agency for Science, Technology & Research (A*STAR). Two Research Councils were formed, then the Joint Council Office to oversee the expansion and integration of research infrastructure for economic development, the launch of an ambitious national scholarship plan to attract and educate 1,000 Singaporean PhD research scholars over 10 years, and the transfer of technology and research manpower to local industry. Infrastructure kept pace with the construction of the iconic Biopolis, then Fusionopolis. Simultaneously, the universities raised its research intensity and built advanced infrastructure to match. The infrastructure houses 18 national research institutes and centres in both the physical and biological sciences and engineering fields. They have attracted and helped many MNC partners to establish their corporate R&D labs and innovation centers in Singapore. In 2006, the National Research Foundation (NRF) and the Research, Innovation and Enterprise Council (RIEC) were established to further expand the scope of research to new areas and lay the foundation for future innovation and enterprise development.

This book traces and discusses the history from national and agency archives and interviews with key people active at various stages of the evolution of our R&D landscape. It also addresses the hows and whys. By understanding the past, we hope the next generation of policy makers and practitioners in Singapore will continue the journey towards building Singapore as an innovation capital in Asia.

The book is organised into nine chapters, with different editors and authors who have the domain knowledge and willingness to spend many hours beyond their official duties to help write and verify the facts and summarise the analysis of decisions made during the various phases of Singapore's research journey. The first chapter, titled *Setting the Stage*, covers the first 25 years of national science and technology development which was necessary for Singapore to partner multinational companies (MNCs) in manufacturing for export. The type of manufacturing industry started as being labour intensive, followed by skills intensive, and then to capital intensive. The critical roles of the Economic Development Board (EDB), the Singapore Institute of Standards and Industrial Research (SISIR) and the Science Council are highlighted.

The second chapter, titled *Shifting Gear into Research*, discusses the findings of the 1986 Economic Committee Report which recommended the strategy of building up long-term research capabilities and capacity for knowledge-based economic development. The translation of that strategy into public policies included the establishment of the National Science and Technology Board (NSTB) which developed the National Technology Plans and convinced the government to provide substantial funding to build the national R&D infrastructure especially the mission-oriented research institutes. Key figures in establishing the infrastructure, role of universities, building up of research culture, manpower challenges, etc were described through stories from interviews of various people in the community. The chapter then covers the renaming of NSTB to the Agency for Science, Technology and Research (A*STAR) with its expanded mission to train local R&D manpower and support the strategic management of research institutes through two research councils. The chapter also describes the achievements of NSTB and A*STAR especially in the building up of national research capacity, the establishment of the National Research Foundation (NRF) and the Research, Innovation and Enterprise council (RIEC).

Chapter 3 describes the multi-agency approach to build up Singapore's R&D ecosystem. It covers the roles of various agencies including NSTB/A*STAR, EDB, NRF, universities, Standards, Productivity and Innovation Board (SPRING), Intellectual Property Office of Singapore (IPOS), Infocomm Development Authority (IDA), Public Utilities Board (PUB), and DSO National Labs. Inter-agency collaboration and public-private collaboration in research are driven by the needs of growing a knowledge-based economy and specific national goals such as the reduction of reliance on water imports. The R&D ecosystem is especially important as Singapore's reputation as a global manufacturing site has to be pivoted on our capabilities to nurture knowledge-based activities such as R&D and innovation.

Chapters 4 and 5 describe the evolution, rationale and roles of research institutes in the physical sciences and engineering, and the biomedical sciences. Key figures who served as champions of the cause, research manpower challenges and difficulties faced are highlighted, followed by the rationale of establishing the two research councils to improve coordination of the research institutes and create synergies within diverse efforts. Vignettes are included to feature some achievements of the respective research institutes and their roles as anchor tenants in Fusionopolis and Biopolis.

Chapter 6 gives an account of the rapid expansion in the research programmes in the two major universities, NUS and NTU, after their early and successful roles in hosting the establishment and development of mission-oriented research institutes of NSTB and then A*STAR. The timing was excellent as the Singapore Government started to plan even longer term by expanding the role of research with the establishment of the NRF. Substantial fundings for new areas are being set aside by NRF to boost the basic research of NUS and NTU since 2006. They include the Research Centres of Excellence which enable selected areas of good university research to become excellent and globally well recognised. A*STAR has also expanded its thematic research programmes and fund universities in emerging science and engineering areas as well as those which will become a pipeline of future technologies upstream of its RIs. This chapter reports on the significance of such efforts to boost the basic research and develop a vibrant research culture in the two major universities and the impressive results to date.

Chapters 7 and 8 together elaborate on the private sector R&D activities, and how public sector agencies and research institutes have played their roles in helping the acceleration of the development of a research network and ecosystem. Successful efforts of EDB and NSTB/A*STAR in capitalising on our R&D infrastructure and human capital to help attract multinational corporations (MNCs) to establish R&D labs in Singapore as their innovation partners are described. The need and examples to help the larger local enterprises and the SMEs in developing R&D competencies to support the MNCs and also to develop long-term indigenous R&D capability and capacity are also presented. The biggest effort to connect scientists with local enterprises is in the 12-year-old GET-Up programme to upgrade the technological capacity of SMEs. The programme details, success stories and recent updates are summarised.

The last chapter covers recent efforts towards Innovation & Entrepreneurship in both public and private sectors. The word "Towards" implies that our Innovation & Entrepreneurship journey is a nascent one and will mature as our R&D capabilities grow. The National Research, Innovation and Enterprise

Plan (formerly called the National Science and Technology Plan) was started only in 2006. The role of public agencies in fostering entrepreneurship is relatively new. The development of a venture capital community, where young people with the desire to succeed as their own bosses exploiting their own Big Ideas, is just recovering from the dot.com bubble burst at the end of the 20th century. More university and RI spin-off companies, fueled by recently introduced public sector innovation grants, nurtured through Incubation Centres or Accelerators, have created a new momentum, especially in the Infocomm Technology space. It is well-known that fostering I&E in Engineering and Biomedical Sciences areas will be more difficult and will take much longer time and more capital investment. But as in the first lap of developing research from scratch, Singapore has the confidence to scale new heights in I&E in a number of areas and build the requisite skills for this new and exciting phase of our knowledge-based economic development.

Chapter 1
Setting the Stage

Hang Chang Chieh and Yeoh Keat Chuan

Fifty years ago, the stage for the development of today's R&D landscape was fledgling. In 1965 when Singapore became independent after a little less than three years in the Federation of Malaysia, the state of science and technology in all sectors was at very low levels. Science and engineering education was basic and limited. Statistics for 1960 show that the total enrolment in educational institutions was 352,952, of which only 8,171 were in universities and colleges, and just 1,257 in technical and commercial institutions. The tertiary institutions were English-medium University of Malaya in Singapore, Chinese-medium Nanyang University, Teachers' Training College for training teachers and Singapore Polytechnic, the latter two set up only in 1954 when Britain began taking steps to divest itself of its Southeast Asian colonies. The University of Malaya in Singapore (later the University of Singapore) had started its Engineering Faculty in 1955 — but in Kuala Lumpur on the premise that the newly set-up campus should have at least one professional degree course given that the medical and law faculties were both in Singapore. The pre-university science classes numbered just eight and prepared students for education in medicine, science and engineering, with the majority going into medicine or science, engineering being lower down the list of desired professions. The medical school had in fact been the first institution of higher learning, starting in 1905, in part because healthcare in the Colony of Singapore was already an important concern. As an open port, Singapore saw (and still does) a constant stream of ships, sailors, migrant workers and travellers who brought diseases that spread easily in the unsanitary and poor housing conditions of those times. Public healthcare had been the spur for the setting up of a unit in the Municipal Government that identified diseases, poisons, drugs and other public health issues circa 1885, evolving subsequently into the Department of Scientific Services. This department eventually evolved into the Forensic Science Division, Singapore's oldest scientific facility.

An open port that has thrived on free trade since the founding of modern Singapore in 1819, Singapore's economic prosperity up to the 1960s during the colonial period was primarily dependent on entrepot trade. In 1959 when Singapore became self-governing, the new government applied to the United Nations for assistance to develop a new economic model for the colonial economy. In 1961, Dr Albert Winsemius, a Dutch economist who had had a hand in reshaping the post-war Dutch economic recovery, led the UN team to advise Singapore on industrialisation. Dr Winsemius would be among the first foreign experts to play a critical role in the development of the new nation and who would remain a consultant to the Singapore Government long after his initial UN assignment was over.

On top of entrepot trade, Singapore in the 1950s had some ongoing manufacturing. In 1960 when Singapore conducted the first-ever comprehensive census of industries employing 10 and more persons, among its findings was that the production/processing of rubber products made up the biggest sector, followed by food industries. Processing of primary products such as copra, tin and timber also contributed to the manufacturing sector. Such a basic level of industrialisation could not absorb the tens of thousands of post-war baby boomers already streaming out of schools by the end of the 1950s. The economic model proposed by the UN team was industrialisation aimed at the expansion and establishment of industries manufacturing goods for the Federation of Malaya and thereafter for export to the world. The Economic Development Board (EDB) started as a department within the Ministry of Finance under Dr Goh Keng Swee. Ngiam Tong Dow who was one of the pioneer EDB officers said in his book, *A Mandarin and the Making of Public Policy: Reflections*: "In the first 10 years of economic development, 1960 to 1970, labour-intensive industries, garments, hair wigs, transistor radio assembly, and ship breaking, saw us through. The label "high tech, low tech" never entered our vocabulary. Any "tech", which could provide our young school-leavers with jobs, would do."

The EDB's job was to attract foreign investors to Singapore and to develop local industries. Once Singapore became part of Malaysia in 1963, this tiny trickle of investors was dammed up by the Malaysian Finance Ministry in Kuala Lumpur. This problem disappeared after Singapore independence on August 9, 1965, but so did the potential of an enlarged domestic base to absorb factory output. One of the attractions of the proposed Federation of Malaysia had been the mooting of a Malaysian Common Market. Thus, Singapore's economic rationale for industrialisation by first relying on the market in the Federation of Malaya before exporting to the world was no longer valid.

Now the world was going to be Singapore's immediate market. However, producing for the world meant products of higher standards and of better quality. They had to be the best they could be to match the quality of the established industrialised nations. Singapore had none of the resources to produce world-class products. The challenges that the nation faced were multifaceted: investors had to be drawn in, appropriate manpower developed, the education system revamped as part of the bigger picture of manpower development, and infrastructure put in place. Thus, as the lead change agent, EDB started out doing everything from investment promotion to financing (Development Bank of Singapore better known today as DBS Bank) to building infrastructure (Jurong Town Corporation better known today as JTC Corporation) to raising industrial standards and handling industry-linked R&D (Singapore Institute of Standards and Industrial Research, today known as SPRING Singapore).

One of the first units formed in the EDB in 1963 was the Industrial Research Unit (IRU) which would be renamed the Singapore Institute of Standards and Industrial Research (SISIR) in 1969 and which remained part of the EDB until 1973 when SISIR became an autonomous statutory board. Started up with Colombo Plan aid from New Zealand, IRU was backed by engineers and scientists from the Singapore Polytechnic and the University of Singapore who undertook research and product development at the specific requests of local industrialists. The EDB's budget of $100 million included provisions for technical and financial assistance to local entrepreneurs. While the IRU was tasked with helping local industries to upgrade and improve their products, another unit, Light Industries Services (LIS), handled a different aspect of manufacturing technology, with both units attached to the Singapore Polytechnic. The differences between the two units were delineated by Finance Minister Dr Goh Keng Swee when he spoke at its 1964 opening at the Singapore Polytechnic: "Briefly, the purpose of the Light Industries Services (LIS) is to help the 2,000 or more light industries in Singapore to expand and to achieve greater efficiency. LIS can help them in five ways, namely through the extension of loans and hire-purchase arrangements for the acquisition of better equipment, through advice on layout, production technique, product designing and standardisation; through management training including training in simple accounting, costing, stores management, guidance in advertising, packaging and marketing, and assistance in getting new sites where needed. However, for more complicated technological problems the LIS has recourse to the Industrial Research Unit (IRU) which is located in this same building. The IRU also has facilities to repair instruments and to carry out tests on products and raw materials." This section of the IRU would in 1972 become the

Instrumentation Systems Centre and one day evolve into the National Metrology Centre. EDB chairman Hon Sui Sen who also spoke at the same opening said: "The Light Industries Services will be working hand in hand with the Industrial Research Unit. They are dedicated not only to serve the existing small industries, but also to help small investors in establishing new ones." The services supplied by IRU included industrial engineering such as plant maintenance, production management and materials handling, general engineering such as plant layout, electrical supply layout and installation, advice on machinery and tools, and industrial design. The unit relied on expertise from international organisations such as International Labour Organisation as well as the academics in Singapore.

Like all Singapore agencies then and now, IRU strove for international recognition as a way of benchmarking performance. In 1966, IRU became a full member of the International Organisation for Standardisation and that same year, it issued its first standard, Singapore Standard on Timber and Primers, an indication of the importance of commodities processing in the industrial landscape. The next year, IRU awarded the first few Quality Certificates to local industrialists whose products met set standards. In 1973 when SISIR became an autonomous statutory board, its founding chairman Dr Lee Kum Tatt said of SISIR: "Its strength lies in the highly trained, experienced professional men and women who provide a corps of technological and managerial expertise from which the Institute provides practical solutions to problems confronting our industries." By 1983, SISIR had issued over 350 standards for industrial and consumer products made in Singapore. Starting by tapping science and engineering staff at Singapore Polytechnic and University of Singapore, it built up its own core of science and engineering manpower to become Singapore's pioneer mission-oriented research institute doing industry-linked research and technology transfer.

Local industrialists known today as Small-Medium Enterprises (SMEs) needed a lot of assistance to improve their products and productivity. The questions that EDB and SISIR dealt with regularly in the 1960s were simple ones as to which standards, which specifications, which properties, what production processes, what properties of raw materials or where to buy production equipment. Although Singapore was the recipient of international standards established by the international organisations, technology transfer, science and technology (S&T) education and training, and dissemination of information were still at rudimentary levels. This was not helped by the fact that the first generation of local entrepreneurs and industry managers were street smart but only minimally educated. Many if not most were family-run enterprises. There was a clearly felt need to promote the importance of S&T in industry but also in education. So enthused was the scientific community about raising public awareness of the importance of S&T in national advancement that several scientific associations were formed in the immediate years following independence: National Academy of Sciences, Science Teachers' Association, and even a National Academy of Chemistry. The key body was the Science Council.

Science Council and its Activities

As early as 1967 the Science Council had been created to push for the national advancement of S&T and to build up human resources in scientific R&D. As defined by the Science Council Act of 1967, the function of the Council was to make reports and recommendations to the Minister on:

- Scientific and technological research and development;
- The effective training and utilisation of scientific and technological manpower in Singapore; and
- The establishment of official relations with other scientific organisations.

The Council was inaugurated on 30th October 1967 in the conference room of the Prime Minister's Office and presided by Dr Toh Chin Chye, then Deputy Prime Minister. Dr Toh was the one member of the Cabinet with a PhD in science and he would become the first Minister for Science and Technology in 1968. In its first annual report for that year, the Science Council reported: "Dr Toh explained that he had tried to make the Council as representative as possible, and had included persons with wide and extensive practical experience. The other six members, who would make up the full council, would be appointed in six months' time so that all areas of scientific interest would be covered." Its inaugural chairman was Dr Lee Kum Tatt who had been one of the first PhDs of the University of Malaya in Singapore, receiving his doctorate in chemistry in 1955. Dr Lee had a very early involvement with scientific services starting with the Department of Scientific Services. Dr Toh pointed out that with limited funds at its disposal, the Council should not be over-ambitious in its programmes at the start but should look into problems which were of immediate and practical value to Singapore. He stressed that the Council should take care that it did not unnecessarily duplicate or interfere with the research or developmental activities of certain departments of the Government, although it was desirable for the Council to help integrate certain of these activities wherever possible.

The Council's basic policy outlined in its first annual report was to confine its attention to "purely practical and utilitarian problems for the next few years although fundamental research especially at institutions of higher learning would not be discouraged". Aware of the high costs of research and to avoid duplication, the Council being made up of men of science saw the "urgent need for better coordination and collaboration of the activities of different scientific bodies in Singapore". The report said: "In this connection, the Council will continue to have a close working relationship with the Economic Development Division of the Ministry of Finance (later to become the Ministry of Trade and Industry), the Economic Development Board as well as other authorities which are planning the social and economic development in the Republic." The Council hoped, too, to get "cooperation between industries and the Government and training institutes in planning an integrated approach to problems relating to scientific, technical and technological education". Plans were drawn up to highlight the importance of S&T, one of which was the creation of a new ministry to achieve better coordination of various S&T programmes. The Council was housed temporarily in the Marine Survey Division offices in Fullerton Building, then given a government bungalow at 23 Nassim Road for its secretariat. It had a full-time secretariat consisting of a secretary and four full-time staff members, all seconded from the civil service. The scientists helming the Council were all part-timers. The Council would relocate to the Science Centre when this building was completed in 1977, and finally to Science Park in Kent Ridge in the 1980s in the last few years of its existence.

Given the general lack of statistical information on a whole range of economic sectors — the first census of industry had been conducted in 1960 — the lack of information on S&T was expected. The Council had noted in its 1967 inaugural annual report: "It is recognised that only after some accurate data are obtained that any projections of future needs can be made with significance." It lamented the difficulty of compiling such information: "The present system of collecting information by questionnaires and correspondence have not been successful. Personal interviews have produced better results and the Council hopes to set up soon a liaison section with full-time staff of its own." The report also lamented its lack of laboratory facilities and its dependence on the laboratories of various government departments. It then proposed the creation of a Ministry of Science and Technology to provide the proper infrastructure to coordinate all the scientific activities in the Republic.

In 1967 the Council took one of the first steps to get the big picture by inviting a UNESCO team consisting of Dr Kurt Billig and D. Merrill to spend six weeks in Singapore to look into the needs and potential for technological research facilities, to give advice on training of technological personnel in the field of metrology and the use of existing equipment, to develop the needs of metrology and the machine tools at the metrology lab of the Singapore Polytechnic, and to look into the possibility of a Technical University of Singapore and the possible establishment of a Technical Documentation Centre. The need for information on what was going on elsewhere and to disseminate such information being considered critical, the Council asked UNESCO for the assistance of a documentation expert, Dr Hans Bauer, to look into the setting up of a scientific-technical documentation centre in Singapore. Dr Bauer's 1969 report, *Proposals for the Setting Up of a Scientific-Technical Information Centre in Singapore*, gives interesting insight into local industry at this time: "In most cases these [local] companies produce products which are sold in the local market. There is seldom any quality control along the production line nor any guaranteed standards of the final product. In order to be able to produce goods which can compete with those from overseas in local or foreign markets, quality control and standards guarantee are essential. Furthermore, in my opinion it will be necessary that these companies must in future start their own technical development in order to improve their products and to keep them up-to-date with regards to design and technical layout. To be able to carry out these aims information is necessary in the form of notification of papers published in periodicals or technical reports, describing new technical processes, new materials, new equipment, new designs, for the specialised industrial branches. If the information is very important, copies of the papers should be made available. In the course of my discussions carried out at many industrial concerns these needs are felt by the young engineers more than the responsible management personnel." This was not surprising given the level of education of the early local entrepreneurs and managers of industry then.

Dr Bauer proposed that the main functions of the proposed information centre would be to provide and disseminate scientific-technical information, assist in solving special technical problems, act as an industrial liaison with manufacturers abroad, and establish close contact with technical organisations, institutions and research centres abroad. Dr Bauer observed at the end of his report: "In my survey under the given terms of reference I have observed that there is a definite lack of scientific-technical information facilities in Singapore. Industries, especially the smaller and indigenous ones and government bodies have expressed the urgent need for scientific-technical information." Dr Bauer's recommendations saw the light of day not in the Science Council but rather in SISIR which was already carrying out some of the proposed functions such as building up its collection of standards as well as assisting local industrialists with their technical problems.

Another step that the Science Council took was the creation of four standing sub-committees: Engineering, Physical Sciences, Social Sciences, and Natural Resources and their Utilisation. The committees were told that their activities "should relate the needs to the interests of the nation so that the research would be purposeful and largely applied in nature. Even more specifically the ultimate aim should be to direct the efforts towards sound development planning through objective knowledge and understanding". Membership of these different standing committees was made up of academics from the tertiary institutions as well as from industry. Thus, the committees were platforms for interaction between industry players and the scientific community. Interaction and building links with the international scientific community being considered important, Dr Lee was included in EDB trade missions to find out for himself what the industrialised countries were doing in the field of S&T. When he became chairman of SISIR, the benefits of such exposure flowed back into SISIR. Among Singapore's earliest interactions with the international scientific community, one was with the International Atomic Energy Agency. Thus, one of Singapore's first scientific conferences was on nuclear energy. In 1968, Singapore was elected to the board of governors of the International Atomic Energy Agency and for many years, there was a unit in the Science Council that dealt with atomic energy matters. This early involvement with atomic energy stemmed from the rapid growth in energy needs in the late 1960s and early 1970s due to rapid industrialisation, prompting the consideration of the nuclear energy option.

In 1968, the Council brought together more than 200 local scientists, technologists, industrialists and economic planners for a *National Conference on Scientific and Technical Co-operation between Industries and Government Bodies* to identify areas of mutual interests and to consider ways and means to foster cooperation. This important conference produced wide-ranging recommendations, among them a national standards authority to prepare and promote Singapore standards, a Research and Development Agency, industrial training for students, interchange of personnel between government bodies and industries to promote better understanding and communication and for optimum deployment of existing manpower. One recommendation that educational institutions orientate their students to industries' needs and that industries provide industrial training and experience for students led to the establishment of the Industrial Training Board to promote and encourage practical training in industrial establishments for students. With support from the Asia Foundation and companies, the Science Council made a quick start by setting up an industrial training fund for about 30 students from tertiary institutions that paid trainees an allowance of about $120 a month. In the first year, some 40 students took part in the industrial orientation scheme during the long vacation in early 1969. The next year, some 550 students took part. With about 100 staff members from the tertiary institutions offering to supervise the attachments, the scheme also gave academic staff an opportunity for a closer interaction with industries "with the view of developing some mutual and meaningful research and development projects". Another recommendation had been for a uniform system of grading trade skills to improve the employment prospects of trade school students. The Industrial Training Board would subsequently come out with the National Trade Certificates. The last recommendation was for the setting up of a Science Endowment Fund with contributions from industries and the Government for the purpose of implementing some of the recommendations.

One of the 1968 recommendations that the Science Council took up was for a study of scientific and technical manpower and to assess future needs. In 1970 a survey of R&D activities in the public sector was conducted, the first such survey. The Council submitted *A Preliminary Report on R&D Activities in Singapore* to the Ministry of Science and Technology. Prepared by Dr Meir Ben Zvi of Israel's National Council for Research and Development, the report consisted of an immediate plan and a long-term plan, probably Singapore's first formal S&T plan. The immediate plan recommended the attracting of foreign science-based industries to set up in Singapore; encouraging big government agencies such as Ministry of Defence, Public Utilities Board, and Port of Singapore Authority to set up their own R&D units; importing talents and experience to guide and lead R&D projects; and giving financial grants to private or public sector agencies with feasible R&D projects. The long-term plan was to develop a sufficient number of qualified and experienced scientific and engineering personnel. One quick-fix would be to send young graduates on two to three years' attachment with industrial establishments overseas. Much of Dr Zvi's recommendations would be remarkably like what Singapore would be doing decades down the road. Some of Dr Zvi's recommendations were already being carried out; others would follow. In 1971, the Defence Minister, Dr Goh Keng Swee, would put together a secret defence science and technology group codenamed Project Magpie. Subsequently named Defence Science Organisation (DSO) and kept secret until the mid-1980s, this would be Singapore's first home-grown R&D body. DSO would one day morph into DSO National Laboratories specialising in defence technology and launch the first Singapore-built satellite in 2011.

One 1968 conference proposal was for the creation of a Research and Development Agency but the Council reported in its 1969/70 annual report that its efforts at this stage were just exploratory. As part of the exploration, the Council set up a Research Grants Committee to vet research project proposals submitted by academic staff in tertiary institutions to the Ministry of Science and Technology. The annual report listed seven research projects all in the agro-fisheries sector. The one that would become a Singapore icon was a project to find a local plant material for the tourist souvenir industry. Said the report: "A study on the possibility of using local plant material for the tourist souvenir industry was looked into. Various methods were experimented with to preserve local plant material so that the colour and appearance could be retained. The results obtained varied with the type of plant material used." The final result of this research project would be the iconic RISIS gold-plated orchids. Dr Lee whose idea the gold-plated orchid was said in an oral history interview in 1993: "Don't forget that to create a national thing takes time. And usually Singaporeans have no time. We want everything instant. How to grow an instant souvenir? … RISIS orchids was a fluke. [We] (s)et up the company on 16 April 1976. Up to 16 April, [the] new company still had no name. So we decided to call it RISIS or SISIR in reverse. That was how RISIS came about." RISIS orchids won the first Best Tourism Souvenir award presented by the Singapore Tourism Board and its continuing popularity as well as its spread to other parts of Southeast Asia is testimony to the brilliance of Dr Lee's idea. The patent for the gold-plated orchid is one of the first patents to originate from Singapore and the company making RISIS orchids is probably Singapore's first successful commercial spin-off from the laboratory. Today, the acronym SISIR has vanished from official circles but it will remain so long as RISIS orchids remain a popular Singapore souvenir.

From the beginning, the Council conducted a wide range of consciousness-raising activities on S&T for young and old. It took up a proposal made at the 1968 conference for the institution of an award for scientists who contributed significantly to the economy of the Republic. The first Gold Medal Award for Applied Research was made in 1969 to Dr TG Ling, an industrial chemist, who pioneered and succeeded in converting the poultry and pig-rearing industry in the Singapore-Malaysia region from "backyard" operations into modern farms using scientifically balanced feeds. The presentation made by the guest of honour, Prime Minister Lee Kuan Yew, was the highlight of a Science Ball which was also the occasion to raise funds for the Science Endowment Fund, yet another idea proposed by the 1968 conferees. The Science Endowment Fund helped the Council to pay for some of its activities as the funding from the Ministry of Science and Technology was used mainly for maintaining its secretariat.

The Council paid particular attention to S&T education of the young from practical programmes to fun activities. The Final Report of the 1962 Commission of Inquiry into Education Singapore had noted that among the factors hampering the expansion of science education in schools were the shortage and the uneven quality of equipment. Science students did not always get a chance to handle even such basic equipment as microscopes, galvanometers or Bunsen burners. In the 1950s and 1960s scientific equipment were all imported and expensive because of strong foreign currencies, and science teachers often kept the equipment locked up carefully or under-used to save on wear and tear and when in use, it was usually shared between many students. In 1970, the Council set up a committee to look into the manufacture of scientific educational kits for students so that they could analyse, experiment and discover scientific facts for themselves. In 1972 the Chinese Chamber of Commerce responded enthusiastically to this project by setting aside $1 million to develop the facility. A year later, the Council worked with SISIR to develop successfully two teaching aids: a student intelligence responder and the electricity and electronics pedagogical aid. The former was to test the response of students to the lessons and the latter came with a book of 62 experiments prepared by two science teachers under the Applied Research Fellowship Scheme. The Council's report for 1974/75 said: "The project has shown that locally-innovated teaching aids have two significant advantages over imported ones — they are tailored to local needs and they are much cheaper. Following up on the project, SISIR has been producing the teaching aids for use in secondary schools in Singapore." In 1975, it organised jointly with the Science Teachers' Association a three-day Science Teaching Aids Competition and Exhibition. It then introduced an award for the most innovative science teaching aids but the contest died out after the initial spurt of interest.

Of more national interest was the Science Quiz organised with Radio and Television Singapore in 1972. The quiz master was Dr Tay Eng Soon who was a lecturer in the Engineering Faculty of the University of Singapore and who was to join DSO as head researcher. Dr Tay would one day enter politics, become Minister of State for Education and introduce the Gifted Education Programme in 1981. Renamed the Science and Industry Quiz, the annual quiz show was so popular with students that the Council published books based on the quiz questions as another way of promoting S&T. There was even a Malay version of the Science Quiz and to raise the ante, two years later the quiz was broadcast "live". In the days before cable television and lively S&T documentaries, the

Council also put energy into developing such documentaries for general broadcast. The Science and Industry Quiz was replaced in 1979 by Young Innovators Quiz which had a shorter run than the original Science and Industry quizzes. Another way of highlighting S&T was in a national exhibition. In 1969 when Singapore celebrated the 150th anniversary of the founding of modern Singapore by Stamford Raffles, together with the Ministry of Science and Technology, the Council mounted an S&T exhibition on the theme of Science in the Service of Man, with the exhibits covering broadly different facets and stages of development in the fields of energy, transportation and communication.

The idea of permanent S&T exhibits had been mooted in the Council's 1968 annual report: a popular science centre "to bring the working of science and technology to the people, particularly the younger generation". Since the beginnings of science, the workings of S&T have always exerted a fascination for young and old everywhere. Back in the late 1950s and early 1960s, schools with pre-university science classes organised public science exhibitions, as did the Science Faculty of the University of Singapore. Such exhibitions of simple scientific processes and experiments were always greeted with oohs and ahs from visitors. The idea of a science museum had come up as early as 1962 when the Final Report of the 1962 Commission of Inquiry into Education Singapore proposed a Science Museum to stimulate interest in the wonders of science. In late 1969, a UNESCO expert, Miss MK Weston of the London Science Museum, arrived in Singapore to advise the Council on the proposed centre. Two years after the initial proposal, the Science Centre Act (Act 33 of 1970) was passed. The Act stated: "It is the duty of the Board to establish and maintain a Science Centre for the purpose of (a) exhibiting objects illustrative of the physical sciences, life sciences, applied sciences, technology and industry; and (b) promoting the dissemination of knowledge in science and technology." Appropriately enough given the importance Singapore attached to linking S&T with industry, the Science Centre was to be located in Jurong right in the heart of Singapore's first industrial estate. The Centre was built with funding from both government and the private sector, such private sector donations even meriting a mention in the newspapers of the day. The Science Centre was opened officially in December 1977 by Dr Toh Chin Chye who, by then, had left the Ministry of Science and Technology to become the Minister for Health.

Ministry of Science and Technology

The importance attached to S&T in Singapore's economic development gave rise to the Ministry of Science and Technology in 1968 but its closure in 1981 was not because S&T was no longer important. Rather the creation of the Ministry had been ahead of its time and in the 1970s its national role had been limited by the size of the existing S&T and R&D landscape. The idea for a Ministry of Science and Technology to give the role of S&T in Singapore more clout had been discussed at the Science Council's inaugural meeting in 1967. Following the early 1968 General Election, the new Cabinet sworn in soon after included the first Minister for Science and Technology, Dr Toh Chin Chye, who was probably one of the first Singaporeans with a PhD in science. Dr Toh had obtained his PhD in Physiology from the National Institute for Medical Research, London, in 1949. His background in science, his standing as a founder-member of the People's Action Party, as well as his drive was expected to give S&T development a big boost. Dr Toh's contribution to the development of S&T in Singapore would be in the area of manpower development and in a re-orientation of the universities towards national needs. As Minister for Science and Technology, his brief was:

- The training and utilisation of scientific and technological manpower at professional level;
- Conducting research into scientific and economic development;
- Coordinating the development of industrialisation and the concomitant specialisation; and
- Making major decisions in the fields of science and technology.

Included in the new Minister's portfolio was charge of the Science Council, Chemistry Services, National Museum, and the Marine Biology Centre. Enforcing the Weights and Measures Act and charge of the Metrication Board and the Industrial Training Board would subsequently be added to the Minister's portfolio.

The development of relevant manpower being the top priority, Dr Toh's first task was the restructuring of S&T education particularly at the professional level. Already the chairman of the Singapore Polytechnic Board of Governors, Dr Toh became Vice-Chancellor of the University of Singapore, a position that had fallen vacant when the incumbent, Dr Lim Tay Boh, died unexpectedly in late 1967. As part of manpower planning, the Ministry published its first *Directory of Scientific and Technical Research and Consultancy Establishments in Singapore* in 1971. Compiled by Wong Hoi Kit and Wee Sin Tho, the introduction noted that the directory excluded medical, architectural and social sciences fields. It was a national survey of S&T activities covering both public and private sectors. Of the 98 establishments and departments listed in the directory, the bulk were in the public sector while of the 13 listed in the private sector, 10 were engineering consultancies and three were laboratories offering services such as assays, analysis and testing of products ranging from oil to ores to foodstuffs. (In the late 1960s and 1970s Singapore was the base for oil exploration in the region, a role it had also played pre-war.)

One of the Ministry's first tasks that had a far-reaching effect throughout all facets of Singapore life was the implementation of metrication. In the late 1960s several of Singapore's key trading partners in Europe were looking into and adopting metric measures as the international standard. Export-oriented Singapore had to do the same to stay in sync with European demands. Pre-metrication, Singapore used a mix of traditional Chinese, Indian and Malay weights and measures such as katis and tahils, gantangs and chupaks, as well as Imperial measures like pounds and ounces, yards, feet and inches. A National Metrication Board was set up end-1970 and metrication programmes intro-duced in 1971. The Ministry also moved to stimulate research by introducing an Applied Research Fellowship Scheme and a Research Grant Scheme, both of which were administered by the Science Council. The Applied Research Fellowship Scheme was industry-linked, while the Research Grant Scheme was intended for researchers in tertiary institutions. In 1975 modifications to the schemes noted that the Research Grant Scheme should fund projects that were more applied in nature "since they arise from Government Ministries inter-ested in using the results". The focus on impact from investments in S&T is one of long-standing.

The 1975 Singapore Yearbook described the activities of the Ministry as:

- Provision of backup scientific services to all government departments for the administration of about 35 Acts and Regulations;
- Administration of the Radiation Protection Act and relevant Regulations;
- Provision of analytical research and consultancy services to commerce and industry; and
- "Conduction" of in-house research and participation in ASEAN regional research projects on proteins.

The same Singapore Yearbook also noted: "Major constraints faced by the Chemistry Dept in meeting the objectives have been the lack of manpower and equipment in the Medicines, Narcotics, Toxicology, Food, Radiation Protection, Microbiology divisions and lack of staff and manpower in the Information and Advisory Services." The Science Council's report for the same year also noted: "Lack of manpower poses the major constraint. There is pres-ently one position for a research officer which is vacant."

Both Ministry of Science and Technology and Science Council lacked more than just manpower in administration or in information and advisory services. The Ministry lacked a strong technical arm despite the several scientific and technical services in its charge and the Science Council had no technical facilities of its own at all. SISIR, the one government body with an S&T research focus, was part of the EDB. In 1973, the year that SISIR was separated from the EDB to become an autonomous statutory board, the Ministry established an Applied Research Corporation (ARC), a non-profit company and an institution of public character with the Minister for Science and Technology as the chairman, and a former EDB Deputy Director, Wan Shoung Fun, as its first managing director. ARC offered SMEs a wide range of services in economic science, business management, engineering systems and industrial R&D by tapping the expertise of the tertiary institutions. The setting-up of ARC stirred the nascent R&D landscape but it could not justify the existence of the ministry. At the end of 1975, in a review of the civil service chaired by Prime Minister Lee Kuan Yew on how best to use the limited number of talented civil servants at its disposal, a review was made of the effectiveness of the Ministry of Science and Technology. The report said: "The role of the Science and Technology Ministry is generally to formulate policies which would decide the optimal use of national resources through the application of science and technology. In the case of developed countries with the means, the Science and Technology Ministry has the added responsibility of deciding policies governing fundamental and pure research. Ours is a nation with no natural resources. Neither can we afford the means to carry out fundamental research. At this stage of development even applied research are not carried out on a scale that requires major policy guidance. With manpower as our only resource, the Science and Technology Ministry is therefore charged with the responsibilities of tertiary technical manpower development, the provision of science and technical services and the promotion of science and technological activities. Even in the area of policy formulation for industrial development, the Ministry plays a very limited role as these functions are largely dominated by the Ministry traditionally in charge. The role and size of the Ministry is such that it inhibits the full development and utilisation of administrative officers appointed to operate it. In the light of the above, it is considered that the Ministry came into existence somewhat prematurely as much of its portfolio could rationally be absorbed by a number of other Ministries." It was then suggested, that if the Ministry was to be retained, its technical arm could be strengthened by adding SISIR to its portfolio.

By the time of the civil service review, Dr Toh had moved to the Ministry of Health. The Ministry of Science and Technology would go through three changes of Ministers — Dr Lee Chiaw Meng (1975–1976), Jek Yeun Thong in 1977 and who was concurrently Ministry of Culture, and lastly, EW Barker from 1978 who was also concurrently Minister for Law and Environment — before closing in 1981. Tertiary institutions and industrial training institutes were taken over by the Ministry of Education while the Science Council, Metrication Board, Weights and Measures Office and the governance of R&D went to the Ministry of Trade and Industry, with the Ministry of Health taking back the Department of Scientific Services.

Singapore Institute of Standards and Industrial Research

The key 1968 National Conference on Scientific and Technical Co-operation Between Industries and Government Bodies had recommended the setting up of a national standards authority to raise the quality of Singapore's manufacturing output. There was already an EDB unit doing just that. In 1968 when the EDB shifted its strategy to export orientation and internationalisation, it was quickly realised that industrial output had to match international standards to succeed. From its basic work of standardisation and quality certification, SISIR's role grew to include testing and verification, non-destructive testing, instrumentation and calibration, industrial design, supplying technical information, extension and technical services, and industrial microbiology. It developed capabilities in engineering metrology, metallurgy and welding technology, production-oriented mechanical design, electroplating and metals' finishing technology, furniture testing and industrial stress analysis. It developed the capabilities that were needed in the industrial sectors important to Singapore at that time.

Starting in the late 1980s, the Institute actively promoted the adoption of ISO 9000 series of quality management standards that were also being adopted by a growing number of industrialised countries, led by the European Community and the USA. Singapore was the first country in the region to align its standards with ISO 9000, adopting it in 1988 and using it as the basis for certifying companies under its Good Manufacturing Practice Scheme that had been started up a year earlier. One important result of the adoption of the ISO 9000 series was that companies so-certified were given access to major export markets without having to spend extra time, money and effort to get separate certification in those countries. SISIR started its international branding with an MOU with the United Kingdom for mutual recognition of each other's ISO 9000 certification schemes. It also signed similar MOUs with seven other leading overseas authorities among them, JMI Institute of Japan,

Deutsche Gesellschaft zur Zertifizierung von Qualitatssicherungssystemen mBh of Germany, and Swiss Association of Quality Assurance Certificates. The number of such MOUs would grow over the years.

SISIR was also growing its R&D facilities under Dr Lee Kum Tatt. As inaugural chairman of the Science Council, Dr Lee had noted that the Council's lack of scientific facilities would hamper R&D activities. When he became chairman of SISIR in 1973, he had the tools to execute his R&D mission more effectively. In SISIR's first annual report, Dr Lee noted: "Virtually all contract R&D and technical investigations undertaken during the last fiscal year arose from specific needs of business and governmental clients." Nevertheless, he could see a more sophisticated kind of R&D being developed and floated the idea of making SISIR a technological institution, saying: "It can use its expertise and facilities to engage in original scientific research, hopefully to make some startling discoveries which may one day prove useful to man. It can also adapt and improve on imported technologies." Thus, he articulated the dream that every true scientist harbours. Being pragmatic, he also said: "SISIR has chosen for the next five years a middle-of-the-line course which is to mobilise scientific and technological knowledge to stimulate and cultivate the indigenous technological base of our industries. The 'mixed technology' approach, that is, the gradual improvement of our indigenous technology by developing and compounding with new foreign technology, was chosen for a number of reasons.... SISIR's aim for the next few years would be to upgrade technology through the provision of adequate facilities, services and technological expertise to industries and to become an effective instrument for technology transfer. It is expected that these will provide the groundwork and climate towards a new sophistication."

Three years later, in the 1976/77 report, he said: "SISIR has also stepped up its activities in research and development. The Institute was involved in developing new products with a view to eventual commercialisation — through contract research, licensing, or royalty arrangements with local entrepreneurs. The objective of this programme is primarily to stimulate greater diversification among local industries while simultaneously fulfilling the Institute's function of promoting, encouraging, and undertaking industrial research for the purpose of developing existing industries or creating new ones. This programme has in fact highlighted the conservative nature of local industries and the lack of research and development capability within their organisations. SISIR will therefore emphasise these developmental aspects of its work and will work closely with local industries to jointly develop products with export potential and which are within the technological levels and skills of Singapore's technocrats. It will be an active policy of SISIR to develop a growing expertise to handle product development projects and to assist in the technology transfer process in Singapore."

SISIR's R&D ambitions grew from strength to strength and in its 1980/81 annual report, Dr Lee said: "Following the explicit Government policy on R&D, SISIR mapped out its 10-year Development and Operational Plan for the 80s." SISIR's emphasis was on industrial R&D with product/process design and development forming the bulk of the R&D projects it handled. In that particular year, the Institute handled a total of 110 projects with fees valued at $1.2 million, a 32% growth in terms of value over the previous year. Plans were made for setting up a Materials Technology and Application Centre with the assistance of an expert from the United Nations Development Programme. The centre would cover the areas of metals, plastics, chemical analyses and electronics materials. The R&D group was expanded with 50 engineers, chemists, food technologists, physicists and metallurgists besides technical support staff. Facilities were also upgraded to include an engineering computer and a computer numerical control machine. It purchased an electron scanning microscope costing $320,000 to cater for the capital-intensive and higher technology industries.

Building a Skilled Workforce

To be attractive to international investors with increasingly sophisticated demands, Singapore needed more than good infrastructure and competitive costs. The country needed a highly-skilled workforce. With the aim to develop this skilled workforce, the EDB also worked on setting up various training institutes in collaboration with MNCs and overseas national bodies. In 1972 the Tata-Government Training Centre was set up, in 1973 Rollei-Government Training Centre, and in 1975 the Philips-Government Training Centre. This skills upgrading initiative was considered so important for the Singapore economy that between 1972 and 1975 male trainees were exempted from National Service (NS). Between 1976 and 1980, male trainees went for their Basic Military Training only after completion of their training centre courses. With the establishment of a body of skilled workers, the NS concessions were removed in 1981. These early training centres linked to specific companies were part of the manpower strategy that had to be implemented at national level if Singapore's economy were to progress. As the industrial environment changed with more sophisticated machinery, with computerised equipment, after the mid-1970s the training institutes had to change too. In 1979 the

Japan-Singapore Training Centre was established to conduct courses on industrial electronics, process and instrumentation control, machinery maintenance, and mould making. All were new skills at the time. In 1983 this centre became the Japan-Singapore Technical Institute (JSTI) that offered two-year courses in industrial electronics engineering and mechatronics. Lin Cheng Ton who was the EDB officer handling the training institutes said in *Heart Work: Stories of How EDB Steered the Singapore Economy from 1961 into the 21st Century*: "We had foreseen that industries would rapidly follow the converging trends of mechanics and electronics and knowledge of mechatronics would become a key competitive advantage for Singapore. We were even told that JSTI was the first in the world to offer an institution-based training programme in mechatronics." The German-Singapore Institute (GSI) opened in 1982 to train "technologists". The term 'technologist" had been coined by EDB officers to describe technicians who could also apply and adapt new technologies during discussions with the Germans on the kind of technical assistance Germany could offer Singapore in setting up a "teaching factory" recounted Lin years later in *Heart Work*. Today the term "technologist" is used by all Singapore polytechnics to describe their technical graduates. Concurrently with talks with the Germans, discussion were going on with the Japanese that led to the establishment of the Japan-Singapore Institute of Software Technology which also opened in 1982. Discussions with the French resulted in 1983 in the French-Singapore Institute (FSI) that was set up to train technologists to focus on electronics. These three institutes were transferred from EDB to Nanyang Polytechnic in 1993.

Besides a skilled workforce, there has to be good supporting industries that form the essential supply chains of the larger companies. To develop more sophisticated industries EDB focused on R&D incentives. In August 1980, *The Straits Times* reported the setting of a new section within EDB to encourage industries to carry out R&D: "The section aims to persuade both multinational corporations and local factories to invest in R and D projects to raise the quality of products. Firms engaged in these activities can apply for financial assistance in the form of double tax deductions for operational expenses, an investment allowance for R and D capital costs and grants provided under the EDB's Product Development Assistance Scheme. The budget for this was $1 million and the maximum granted to applicants was $100,000 for each project." The head of the new unit was Dr Vincent Yip who said then that the potential for industrial R&D lay in design and custom work, engineering and consultancy services.

In 1978, the Government introduced the R&D Block Vote that in 1983 became the Research and Development Assistance Scheme (RDAS) administered in part by the Science Council. The Council was also responsible for the publication of Project Summaries of the RDAS projects and regularly organised technical talks and seminars that were presented by Principal Investigators (PIs) and open to the public in order to encourage direct communication with the PIs. Both public sector organisations and private companies (MNCs and local enterprises) were eligible to apply for the RDAS grant. By the end of 1989, RDAS projects numbered 85 with total fund commitment of $56.5 million. The project research areas were either in the Engineering and Physical Sciences or the Biological and Medical Sciences. A number of multinational corporations (MNCs), public agencies and local companies, for example, Seagate Technology, Telecoms Authority of Singapore, Chartered Industries of Singapore, and Pacific Biomedical Enterprises Pte Ltd, benefitted from the sizeable grants and many took the opportunity to initiate joint R&D with universities and other public agencies. Often in collaboration with industry partners, the relatively large grants enabled academic staff in National University of Singapore (NUS) and Nanyang Technological Institute (NTI, predecessor institute of Nanyang Technological University) to do larger scale research. One noteworthy project that emerged from the RDAS scheme project was the Microprocessor Applications Centre (MAC) undertaken by Applied Research Corporation in collaboration with NUS. The Centre was set up as an applied R&D centre which also undertook turnkey automation projects using the new microprocessor technology and developed microprocessor-based products for commercial and industrial applications for industrial clients. In addition, the Centre provided training courses, seminars, technical assistance and consultancy services in the emerging microprocessor technology for local industry. With the support of three consecutive RDAS grants from 1982 to 1988, MAC developed capabilities in advanced technologies such as machine vision and control, distributed processing and multi-processing. In August 1988, the Centre was transferred to SISIR to synergise with SISIR's existing centres of competence.

This sharpening of the focus on R&D from 1981 came at the same time as the closure of the Ministry of Science and Technology and a reduced role for the Science Council which came under the purview of the Ministry of Education, its original statutory role of advising the Government on S&T matters having narrowed to mainly promotional activities, establishing links with international S&T community and agencies, and organising seminars and conferences. The science and technology restructuring in progress affected SISIR's view of its role. In its 1982/83 annual report, Dr Lee stated that the Institute had been restructured to place greater emphasis on its core programmes of Quality, Quality Assurance and Metrology and the sub-heading "Industrial Research and Development" disappeared from the annual report. However, the next year (1983/84), "Special Projects" appeared and in 1984/85 there was "Innovation Projects". This was Dr Lee's last SISIR annual report and he re-introduced R&D projects into it, saying: "(W)hile the promotion of quality and reliability, testing, quality certification and consultancy remained as the important mainstay of SISIR's programmes, innovation, R&D work, and the development of new methods and capabilities to meet industries' needs provided the challenge for staff. An increasing number of such R&D projects were embarked upon in 1984. Several of these projects were undertaken in collaboration with a number of identified 'model' companies which have demonstrated a commitment to quality." Dr Lee had always believed that men of science needed a "blue rose" to strive for to achieve excellence. In a 1993 oral history interview, he said: "If you are a true scientist, I have to appeal to you emotionally. You need research. If not, you are not a scientist…. Up to today, I am going to promote my blue rose. You have something to live for. We all have." He had a vision for SISIR that explains why it came to take the lead in R&D in the 1970s and 1980s. He said: "So I decided to build SISIR. SISIR is our DSIR [Britain's Department of Science and Industrial Research] in a small way. … I decided that I should not do creative research — leave that to the university — but to do innovative research. … What is the difference between creativity and innovation? I'll tell you what is innovation. Innovation is, you take somebody else's work and improve on it. Then if it is improved, it's innovation. If it is the same, you copied. If it's worse, you lose your money. The Japanese are very good at that. But creativity means you do something out of nothing. It's not there before and you promote it. That's creativity. It's more hard. So I decided that we must do innovation."

By 1987 when SISIR moved to its new premises at 1 Science Park in Kent Ridge, it had built up its scientific and technological credibility to such a degree that it was recognised as an inspection agent by some 10 leading standards institutions in developed countries such as Canadian Standards Association, British Standards Institution, Germany's Technical Inspection Organisation, and Japan's Ministry of Trade and Industry. Its test reports were recognised by some six regulators including the US Food and Drug Administration. It set up a Product and Process Technology Centre to support product and process development as the basis for innovation and technology upgrading by Singapore manufacturers. It expanded its capabilities in food technology into a Food Technology Centre to undertake food technology research and development and provide technical consultancy services to traditional food manufacturers to help upgrade their products and expand their product range. Its capabilities included food processing technology, packaging and preservation technology, product formulation, microbiological and chemical analyses, and food plant engineering with certain Singapore foods presenting interesting challenges. At its new premises at Science Park, the Institute also included among its facilities a Prototype Development Centre to assist industry in R&D, incorporating special industrial research units for industrial personnel to carry out specific R&D projects, called the Industrial R&D Incubator Centre. There was an expanded library with the Institute's extensive collection of industrial standards from around the world as well as a patent information service to keep innovators in touch with the latest patent-related developments worldwide. The facilities at 1 Science Park included metrology laboratories which must be purpose-built to maximise accuracy of its measurements. (The laboratories remain in use today as part of the National Metrology Centre which became an A*STAR institution in 2008.) Its 1986/87 annual report gave SISIR's mission as: "To promote and upgrade quality and technology in the manufacturing sector, with the aim of enhancing the international competitiveness of Singapore-made products." Its chairman Col. (Res) Quek Poh Huat said: "We strove to remain in the forefront of Singapore's industrial research and development by embarking on several new programmes which ensured our continued ability to meet the increasingly sophisticated needs of industry." In 1986, it had introduced a Singapore Laboratory Accreditation Scheme to assess and accredit Singapore laboratories to internationally accepted standards. The scheme also offered technical assistance to private laboratories to upgrade their facilities. At the same time, the SISIR laboratories acted as back-up technical support to complement their services. This value-added service sector had been identified in the 1986 Economic Committee as a growth sector with export potential.

In 1987 SISIR's technical competence centres included:

- Design and Development Centre to collaborate with industries in new product development;
- Materials Technology Centre to help in the selection, evaluation and application of industrial materials;
- Food Technology Centre for the development of new food products and processes;
- Metrology Centre to maintain the national physical standards of measures such as length, mass and electrical parameters;

- Industrial R&D Incubator Centre to support SMEs keen to start an R&D facility; and
- A Patent Information Service located in its technical information and documentation library.

During the 1980s while SISIR was forging ahead, several attempts were made to beef up the Science Council's role in S&T matters. Its budget was increased. In 1984 it got its first full-time executive director, Dr Vincent Yip, who had previously handled the EDB's Product Development Assistance Scheme (PDAS) that continued to run for a number of years in tandem with the multi-million-dollar RDAS. The Council's secretariat was moved from Science Centre to Science Park in 1987. Moving the Council out of the Science Centre was an attempt to change public perception of the Science Council as an educational body which it was never intended to be, unlike the Science Centre which had been intended as an educational institution and thus came under the purview of the Ministry of Education. Following the move to the Science Park, the Council took over responsibility for two major sets of S&T activities: the administration and promotion of RDAS, and the administration of Science Park. The beefing up of the activities of the Science Council was part of the heightening awareness of the importance of S&T in the modern world. The awareness had been sharpened by the arrival of MNCs exploiting the latest technology and the latest discoveries in biomedical sciences — all the result of the hard work of EDB officers who combed the developed world in search of MNCs to invest in Singapore's economy. In 1981 Apple set up in Singapore, initially to produce circuit boards but quickly moved on to producing Apple II computers with parts sourced in Singapore. Apple's presence attracted computer peripherals manufacturers, soon drawing in the inventor of the hard disk drive, Seagate Technology, which set up manufacturing facilities in 1982. Two years later, Seagate had also set up its own R&D facilities in Science Park. In 1984, SGS Singapore, a subsidiary of Italian semiconductor manufacturer SGS Microelecttronica, opened Singapore's first semiconductor wafer diffusion plant in Ang Mo Kio Industrial Park 2, a plant that had taken several years of EDB work to bring into Singapore. At the same time as the electronics manufacturing sector was moving beyond TV sets and small appliances, the biomedical sector was stirring. In 1982 Glaxo had set up a plant to manufacture the active compound, Rantidine Hydrochloride, for its new drug, Zantac, discovered in the 1970s. The Singapore Glaxo plant was the first plant here to get a US Food and Drug Administration inspection for which it got a clean bill of health. Years later, Zantac was to be rated as one of the world's most profitable drugs and Glaxo as one of the 10 most profitable pharmaceutical companies. That year, too, the Department of Obstetrics and Gynaecology of the NUS Faculty of Medicine had done Asia's first successful in-vitro fertilisation and embryo transfer, with the baby boy being born on 19 May 1983. This was also the year that Petrochemical Corporation of Singapore (Sumitomo Chemical) and its first downstream affiliates — Philips Petroleum Singapore Chemicals, Polyolefin Company (Singapore) and Denka Singapore — started up plants on Pulau Ayer Merbau which would one day become part of Jurong Island. While Singapore had been a pre-war oil refining centre, downstream petrochemical industries such as plastics were new. That same year, 1983, Science Park was officially opened.

In fact, the 1982 Singapore Yearbook had declared: "The prime objective of the Plan (Singapore's Economic Development Plan for the Eighties) is to develop Singapore into a modern industrial economy based on science, technology, skills and knowledge." In 1982 at the official opening of Glaxochem Pte Ltd, the Minister for Trade and Industry, Dr Tony Tan, was reported in *The Straits Times* as saying: "It may be too early for us to build the bridge between the present and the future when biotechnology has a major impact on society, but it is certainly not too early for us now to plan and to learn how to cross the bridge when the time comes." Emerging technologies and industry were racing far ahead of tertiary teaching institutions such as polytechnics and universities and the presence of training institutes such as German-Singapore Training Institute and French-Singapore Training Institute impressed upon investors in emerging technologies outside Singapore that the country's economic strategy was to commit not just to the established but also to continuously innovate and stay ahead with new technologies. If the economic strategy in the 1970s was to focus on skill-intensive industries and services, the strategy in the 1980s moved towards focusing on capital-intensive industries and services. It was none too soon. In the early 1980s several external and internal factors caused the slowing down of Singapore's economy and brought about the first post-independence recession in the mid-1980s. It prompted the setting up of a high-level economic review committee headed by the Minister for Trade and Industry in 1985. The publication in 1986 of the Report of the Economic Committee (EC) titled *The Singapore Economy: New Directions* would change the whole R&D landscape beyond recognition.

Chapter 2

Shifting Gear into Research

Hang Chang Chieh, Low Teck Seng, and Yeoh Keat Chuan

After 1965, Singapore's economy had expanded rapidly notwithstanding the 1970s oil crisis and downturns elsewhere in the world. Even in a bad year, GDP growth was about 5% while in a boom year it stood at about 15%. Between 1980 and 1984, real GDP growth averaged 8.5%. The Economic Committee which was set up in 1985 to review the economy published its report titled *The Singapore Economy: New Directions* in February 1986. It concluded: "The period of easy growth is now over. The recession of 1985 and 1986 is a turning point in our economic development. Even after we overcome this recession, economic growth will not rebound to its previous average of 9%. Internally we have finished doing all the easy things that can foster growth. Externally, international trade is no longer expanding as exuberantly as it used to, and worse, the trading environment is becoming increasingly hostile. From now on growth will be harder to achieve."[1] The report identified some internal factors that contributed to the slowdown. One was the high wage policy of the late 1970s. The policy had been instituted by the National Wages Council (NWC), a tripartite body made up of employers, trade unions and the Government, and formed in 1972 to formulate wage guidelines. NWC had been formed at a time when Singapore was industrialising rapidly and facing a growing labour crunch. Raising wages was one way to attract more people into the work force. The higher wages cut into Singapore's cost competitiveness at a time when regional economies with much cheaper and more plentiful labour were competing for these jobs. Other operating costs — rentals, interest costs, statutory board charges — also went up, further decreasing Singapore's cost competitiveness. The Report contained detailed changes to specific policies but the four key recommendations were wage reform, tax reform, upgrading of business efficiency and productivity, and promotion of services. Manufacturing had to remain a core component of Singapore's economy as it would provide a

---

[1] Singapore. Economic Committee Report. The Singapore Economy: New Directions. Singapore, Ministry of Trade and Industry, 1986; p. 4.

wide range of employment opportunities. The Report noted: "Since 1983, strong expansion in new high value-added industries such as computers, electronics, machinery, printing and pharmaceuticals provided the impetus for growth. This was achieved through a high rate of new and better quality investments in these industries." The Report also noted that the high GDP growth of the first half of the 1980s was generated largely by the strong growth of international services which accounted for 33% of GDP or about 1.5 times the manufacturing share. Productivity growth in this sector averaged 8.5% per annum. "The present level of value-added per worker in these services, $48,000 in 1980 factor cost, is twice the national level." It was clear that Singapore's economic restructuring had to focus on adding value to the different sectors, and developing the capacity to add value. One source of higher value-added would be higher productivity. The Report said: "Analysis of our performance in the last two decades shows that productivity growth depends on three main factors:

- Increases in capital-intensity, as measured by the rate of growth of investments in machinery and equipment per worker;
- Increases in the proportion of professional and technically-skilled workers engaged in knowledge-intensive occupations; and
- 'Technical progress', i.e. qualitative improvements in the overall efficiency of our businesses."

The Report said: "Since 1980, the Government has placed much emphasis on high-technology. It recognised that Singapore must keep abreast with the rapid advances in new technologies and move into the emerging new generation of high-tech industries, as they are the growth areas of the future. To continue depending solely on established industries, such as oil-refining and shipbuilding and repair, which had been growth industries of the late sixties and seventies, would eventually lead to decline, industrial stagnation and unemployment. While the need for a high-technology policy is obvious, we should clarify our goals and objectives. Our goal should be to achieve higher value-added content, not high technology *per se*. High-tech activities represent only one area for future growth. But the emerging advanced technologies will have a significant impact even on established industries. In some cases, production processes will be radically altered, while in others, new patterns of demand will emerge. All industries, both in manufacturing and services, will benefit by upgrading and exploiting new technologies."[2]

---

[2] Singapore. Economic Committee Report. The Singapore Economy: New Directions. Singapore, Ministry of Trade and Industry, 1986; p. 147.

Philip Yeo who was the Chairman of EDB in 1986 recalled in an interview in 2011 for the A*STAR commemorative publication: "Singapore was making TVs, radios, irons — Philips irons — making products, assembling it. So I said No. The focus for us will be components — disk drives, semiconductors, chemicals, pharmaceutical chemicals — which can be exported anywhere. I decided that we must do R&D or product development. I was trying to get the economy going in the 1985–86 recession when a lot of people had lost faith in manufacturing."[3] The restructured industrial policy would aim to attract the leading MNCs in the new growth industries of the future. EDB would also work to attract young foreign technology companies with the potential to become MNCs.

The Economic Committee Report stated:
The goals of the high-technology policy should be:

- To encourage all industries to exploit and apply new advances in technology as widely as possible;
- To draw up specific plans to develop competence in selected new technologies; and
- To move into high-technology industries as one area for growth.

The high-technology policy outlined in the report was a far-sighted plan to restructure the Singapore economy with advanced manufacturing and R&D capabilities. It was a bold move but it was also a fairly tall order given that there was no well-established tradition of research in tertiary institutions which were viewed primarily as teaching institutions. Research was not a priority, and the annual graduation of PhDs in science and engineering a mere handful. However, there was a lively project-based applied research scene set up by the largest public sector R&D organisation then, SISIR, which tapped the expertise in the universities and polytechnics to handle research for industry clients. There was also the Science Council which since its formation had been promoting science and technology in a variety of ways but which became more active in its economic mission after the appointment of a full-time director in 1984. By the time of the 1986 Economic Committee Report the R&D landscape included several institutes attached to the National University of Singapore, the one existing university until 1991. In the 1980s, another public sector prime mover in engineering research was the Ministry of Defence (MINDEF) which was focusing on developing domestic defence technology capabilities and

---

[3] A*STAR: 20 Years of Science and Technology in Singapore. Published by Agency for Science, Technology and Research, 2011; p. 38.

whose investment in this specialised R&D was kept under wraps until years later. (DSO National Laboratories and Defence Science Technology Agency were unveiled in 1997 and 2000, respectively.)

In 1986, Singapore did not have a coordinated R&D plan. There was also not much experience in R&D management although a tiny core was developing because of the defence technology R&D. Thus, some of the personalities in that arena came to be involved in the national R&D plans: Teo Ming Kian, Philip Yeo, and some of the engineers working on defence technology, many of whom were university academics who were awarded MINDEF research contracts and who would also serve in various capacities in the R&D landscape after 1991. There was a small body of PhDs that had been building up since independence in 1965 when Singapore's education system began investing heavily in science and engineering. Singapore's economy was moving into a different phase. In the 60s, it had been labour-intensive; in the 70s, skills-intensive; in the 80s, it was capital-intensive. Going into the 21st century, it had to be knowledge-intensive. The efforts in applied R&D in the 1970s gave the planners some confidence in Singapore's ability to carry out incremental product development. SISIR had successfully driven the quality of many Singapore-made products up to international standards. But could Singapore pull together its scarce knowledge resources and rapidly build new knowledge to make a quantum leap into a more knowledge-intensive economy? To bring this about, Singapore had to put in place an effective R&D policy that would build up its knowledge resources which naturally had to include developing first-rate scientific and engineering manpower. Although at that time there were talented and highly-motivated individuals who were engaging in basic research, the official route to developing the R&D landscape had to be through link-ups with established and emerging technologies brought to Singapore by MNCs in the 1980s. Thus the EC Report identified the emerging technologies and new growth industries that had been setting in Singapore in the 1980s. Among these were communications and information technology, pharmaceuticals, robotics and artificial intelligence, microelectronics, laser technology and optics, and chemicals.

Start-Up Strategies

The proposed R&D strategy had four components: tax incentives, infrastructure and manpower development, and funding. The most straightforward were the tax incentives as most of Singapore's R&D activities were being undertaken by private enterprises. In addition to the existing (1986) tax incentives for

R&D, viz. 50% investment allowances for R&D capital equipment and double deduction for R&D expenses, a proposal for tax deferral for R&D reserves was made. This tax deferral scheme would allow approved companies to set aside 20% of their profits as R&D reserves, to be given tax exemption if spent within a period of three years, thus giving companies an additional incentive to engage in R&D.

The second and third component of the proposed R&D strategy, that of infrastructure and manpower development, are interlinked, crucial and far more complex longer-term challenges. The PhD output from the one university was miniscule and the bulk of PhD education was done in overseas universities with the majority of PhD graduates returning to become academics. In fact, there was little encouragement of PhD education. If one got down to the basics, there was very little encouragement of curiosity and creativity in the primary and secondary schools. Young people were given few opportunities to tinker, invent and use their hands in a Singaporean culture that in general fostered conformity and discouraged the asking of questions. Beyond this more general cultivation of scientific curiosity among young Singaporeans as a long-term strategy to develop knowledge workers as well as scientific and engineering talent, there was the more immediate need to expand post-graduate education in science and engineering to develop the pool of Research Scientists and Engineers (RSEs). In 1990, on the eve of the formation of the National Science and Technology Board (NSTB), there were 28 researchers per 10,000 workers.[4] The number of PhD researchers (including those in the universities) stood at 970.[5] A thriving R&D ecosystem had to have a much higher ratio of researchers to workers and there should be a greater number of PhD researchers. There was also the need to raise the social status of RSEs. Because there was no tradition of research, the brightest students went into the civil service and the larger private sector corporations, almost never into PhD programmes and research. (In the 21st century, banking and financial services would prove to be serious competitors for scientific and engineering talent.) Teo Ming Kian, former Chairman of NSTB, recalled in an oral history interview in 2014: "I went to the universities to try and convince the students to take on R&D as a career and the first question after my presentation was 'What's so exciting about being a lecturer?' It's telling that at that time they equated R&D as lecturing. That's how they saw it. That's the only people they saw doing R&D. It's almost like when

---

[4] A*STAR: 20 Years of Science and Technology in Singapore. Published by Agency for Science, Technology and Research, 2011; p. 113.
[5] A*STAR: 20 Years of Science and Technology in Singapore. Published by Agency for Science, Technology and Research, 2011; p. 113.

I was in MINDEF trying to build up our defence technology group. Went to the universities several years back, and again after the presentations they said 'Why do you need defence engineers, carrying rifles to charge up the hill?'" Thus, the task of developing RSE resources also involved changing basic mindsets — and creating the career opportunities for RSEs. Apart from becoming an academic, someone with a PhD could not find a career in R&D back in the 1980s.

Besides developing scientific and engineering talent, Singapore's nascent R&D ecosystem also needed infrastructure and an institutional environment that would support R&D activities. Apart from facilities in NUS, existing infrastructure consisted essentially of the Science Park. The budget for this was allocated in 1980 and the Park officially opened in 1984. While the establishment of Science Park in Kent Ridge in the mid-1980s was a step in the right direction in bringing together companies engaged in emerging technologies and carrying out their own R&D, this was insufficient, as was the Science Park's mere physical proximity to NUS in Kent Ridge. Institutional links were weak in the 1980s and there was the need to encourage greater university-industry interactions. Although the Science Council and SISIR were generating research interaction through their various projects, it could be a lot more impactful and it could be a lot more focused. Not only were institutional links weak, the institutions engaged in R&D were either small, university-based or geared basically for training and teaching. What Singapore needed were research centres of competence in areas of relevance to the identified growth industries. During this time SISIR ran the Materials Technology Applications Centre, Microprocessor Applications Centre, Food Technology Centre, Design and Product Development Centre, and it had an Industrial R&D Incubator Centre to support SMEs keen to look into setting up R&D facilities. In NUS, there was the Institute of Systems Science (ISS); and the National Computer Board (NCB) which had been set up in 1981, started its own research institute, Information Technology Institute (ITI), in 1986. These institutes were beginning to try their hand at R&D but were still focused on training and constrained by insufficient funding. The piecemeal client-funded research going on in the 1980s was too fractured to make any big impact although it should be noted that even at this stage, companies that engaged in some kind of R&D were better able to develop new product lines and expand compared to companies that did not. One significant industrial sector that did well by turning to R&D was Singapore's food industry. Through R&D particularly at Singapore Polytechnic, traditional manufacturers of household staples such as soya and chilli sauces and noodles were able to expand and export food products that met international standards. Today, some of these food products are arguably amongst the best in the world. (One food product that appears to have captured the world market is popiah skin. It is to be found practically all around the world.[6]) To similarly boost the identified growth industry sectors, the R&D had to be intensified in these emerging technologies. More institutes of advanced engineering had to be set up. This necessity had been highlighted in a Development Project Committee (DPC) Paper prepared in 1989 which grew out of a high-level meeting chaired by then Minister for Education, Dr Tony Tan, a scientist by training and a PhD graduate of the University of Adelaide, and a long-time supporter of university research.

This meeting on 31 October 1989 was held to discuss ways and means of promoting and developing R&D. Present at this meeting were representatives from the Ministry of Trade and Industry (MTI), NUS, NTI and EDB. The meeting's agenda was wide-ranging, covering aspects such as manpower training, staffing of the institutes, and ways to accelerate R&D activities in Singapore. Following the meeting, a DPC paper was written up by Prof CC Hang and Prof Brian Lee, of the NUS Engineering Faculty and Engineering School of NTI respectively, and submitted to the Ministry of Finance (MOF). It outlined the following goals for the Institutes of Advanced Engineering to be established in Singapore:

- To undertake R&D and develop Centres of Excellence in selected areas of advanced engineering research relevant to the critical needs of Singapore industry;
- To train and develop a pool of industry-oriented researchers because the universities do not always train industry-ready researchers;
- To participate in joint research programmes and provide technical assistance and consultancy to help Singapore firms to develop; and
- To develop effective mechanisms for rapid transfer to industry of research findings and technology generated in the institutes.

In June 1990, MOF approved the plan with a three-year budget of $81.56 million for Phase 1 development of the first two Institutes of Advanced Engineering: Institute of Microelectronics and the Institute of Manufacturing Technology. (This second institute was to be merged with GINTIC Institute of Manufacturing Technology (originally Grumman International NTI CAD/CAM Centre (GINTIC)), to form the Singapore Institute of Manufacturing Technology or SIMTech.) These Institutes of Advanced Engineering were to report to a National Council for Science and Technology to be set up under MTI.

---

[6]Research by Lee Geok Boi during her travels around the world.

From Science Council to National Science and Technology Board

After the 1986 Economic Report and the highlighted focus on S&T as an important part of Singapore's economic restructuring, moves were made to strengthen the Science Council profile in order for it to tackle the burgeoning task of building up an R&D ecosystem. The Council took over responsibility for two major sets of S&T activities: the administration and promotion of Research and Development Assistance Scheme (RDAS), and the administration of Science Park. Now that Singapore's S&T needs were of different and higher levels, the Council had to undergo a serious makeover from its mostly advisory roles if it were to be the agency for leveraging R&D to meet changing economic needs. It was easier said than done as the Council had only four full-time staff.[7] The Council had been led by part-time academics for much of its life. The new National Science & Technology Board (NSTB) would take over the functions of the Science Council and start with staff of 18.

In 1989, while in Tokyo for a meeting and over a dinner of porridge with then Minister for Trade and Industry BG (Res) Lee Hsien Loong, then EDB Chairman Yeo suggested replacing the Science Council with a science and technology board. Yeo recounted for the A*STAR commemorative book the points he made at that time: "If we just had manufacturing alone without R&D, we would not be able to develop new products. So I said, 'we must co-locate R&D and manufacturing under one umbrella.'"

Former NSTB Chairman Teo Ming Kian elaborated, also in the same book: "For Singapore, it was imperative for us to do so or we would lose our competitive edge to be a place for companies, both Singaporean and foreign, to start up, stay and grow. And this required a capability and capacity for knowledge creation and innovation to translate the new knowledge for the marketplace. Having a growing pool of researchers and a credible research capability in our economy, both in the private and public sector, was therefore crucial to fire up the creation of a vibrant and sustaining science and technology environment for an ecosystem of large, small, local and foreign enterprises to develop."

The plans for a new science and technology board were announced in August 1990 by the Minister for Trade and Industry, BG (Res) Lee Hsien Loong: The Science Council was to be upgraded to a statutory board with a projected initial budget of $100 million to develop Singapore into a research hub to boost competitiveness in selected industry and services sectors. The mission of the new board was to identify promising new fields of research while strengthening

---

[7] Recalled by Lee Ying Adams in a 2015 oral history interview. She was a chemist with SISIR who joined NSTB as an administrator when it was formed in 1991.

existing R&D activities, and to identify and nurture manpower needs to meet the demands of R&D work in Singapore. Originally called National Council for Science and Technology,[8] it was finally renamed National Science and Technology Board (NSTB) with Lam Chuan Leong as the founding chairman. He was also then Permanent Secretary of the Ministry of Trade and Industry. NSTB began operations on 11 January 1991 with the announcement of its role of promoting and funding R&D projects in public and private sectors, planning for future research institutes (RIs), and the formation of two new institutes. In line with its mission of developing Singapore into a hub of excellence in selected fields of science and technology to enhance national competitiveness in the industrial and services sectors, NSTB charted the first National Technology Plan to set directions for the development of core competencies in selected new technologies (including advanced manufacturing technologies) which could result in national comparative and competitive advantages.

National Technology Plans

The first five-year National Technology Plan (NTP) had an allocated budget of $2 billion to develop S&T in Singapore. The Plan was produced in consultation with 200 experts grouped into nine subcommittees from both the private and public sectors. Published in August 1991, a few months after the formation of NSTB, the National Technology Plan took as its base the first National R&D Survey done by the Science Council in 1978. Investment in R&D then was 0.2% of GDP. The long-term goal was to raise Singapore's R&D expenditure to 3% of GDP while the short-term goal was to reach 2% as soon as possible.

The Plan outlined Singapore's strategy of concentrating on research that would lead to economic upgrading. NSTB's inaugural Chairman Lam Chuan Leong (1991–1993) said in A*STAR's 2011 commemorative publication: "The broad direction of harnessing R&D to serve economic growth was reasonably clear. The sectoral emphasis was also reasonably clear. The key issue was where to position precisely the R&D effort along the spectrum from more upstream to more downstream. Another issue was how to justify the money to be put into R&D and how to measure the benefits and over what period of time. Over time, the faith in R&D has increased but in the initial phases, such questions were frequently posed and I think we did not know exactly how to go about answering them."

---

[8] Business Times, 5 Oct 1990, p. 2.

His successor Teo Ming Kian (1993–2001) said for the same publication: "It was mandated by the Ministry of Trade and Industry and the Ministry of Finance for NSTB to support and fund applied research with economic payoff. Given this direction, it was quite obvious that the fields of R&D focus were in our industry sectors like electronics, engineering, petrochemicals and info-communication. To ensure good alignment with this focus, NSTB was reorganised with departments specifically looking after these sectors with a clear objective of enhancing our economic competitiveness. The objective was three-fold. Firstly, to leverage on R&D to strengthen the capabilities of companies and help move them up the value-added curve so they could be more competitive. Secondly, to help transform the industry. Thirdly, to pre-position ourselves to take advantage of opportunities when they arise even though the industry had not yet been developed."

By 1991, some fields were clearly maturing and vital to the economy: medical sciences, electronics, energy, water environment and resources, biotechnology, manufacturing technology, information technology, materials technology, chemicals. However, the initial plans were kept broad to serve as a guide and to give latitude for development as well as to capture new opportunities. Given the high cost of cutting-edge research and the risks involved, Singapore chose to focus on niche research areas where, by the turn of the century, their R&D efforts might begin to have an economic impact.

NSTB worked closely with EDB to understand the needs of companies in the different industry sectors. What the different industry sectors needed by way of R&D differed depending on whether they were local SMEs or MNCs. Many of the MNCs in Singapore in the 1980s, for example, Motorola, Apple, Glaxo, Seagate had in-house R&D centres located in their home bases in the US or UK. Meeting their industry needs required a different strategy from one for the local SMEs whose experience with R&D was virtually non-existent.

For the MNCs, globalisation was shaping up as the dominant business trend in the 1990s. The EDB had also embarked on an effort to persuade MNCs to set up their regional headquarters in Singapore, a strategy of growing the services sector that had been identified in the 1986 Economic Committee Report. With the regional base headquartered in Singapore, MNCs would be closer to their Asian clients, be able to understand, respond and meet the needs of a rapidly developing Asia, thus making themselves more competitive and more cost-effective compared to competitors based further away. At the same time, in support of the R&D plan, EDB encouraged MNCs to set up R&D centres in Singapore. The strategic aim was to deepen the roots of MNCs in Singapore.

Singapore's economic restructuring to leverage on S&T was in fact coming at a time when the idea of R&D internationalisation was beginning to take shape. Although the term "Open Innovation" would not be coined until the start of the 21st century,[9] essentially this was what Singapore's economic restructuring strategy that focused on developing international services hoped to achieve. Even without this trend to boost the development of the R&D ecosystem, Singapore had to be prepared to support both advanced manufacturing and R&D centres to be hosted here to enhance its economic competitiveness. With rising wages and insufficient compensatory productivity gains in the 1980s, Singapore had lost the competitive edge of low-costs industries. To be attractive and retain MNCs, Singapore had to add value to manufacturing which included co-locating R&D.

For the MNCs, there were essentially two types of foreign R&D sites: the "home-base-augmenting" site to tap knowledge from competitors and universities around the globe, and the "home-base-exploiting" site to adapt standard products to local and regional demands. Both types of R&D activities were welcomed by Singapore to further promote and reinforce the long-term partnership of MNCs with local institutions. Singapore therefore decided to establish relevant mission-oriented research institutes to develop core competencies ahead of time so as to allow them to reach a critical mass to perform their specific roles effectively. The roles of these institutes (as planned in the Institutes of Advanced Engineering proposed in the 1990 DPC paper) in supporting the local R&D needs of MNCs would include:

- Recruiting and nurturing a "float" pool of Research Scientists and Engineers (RSEs) who are competent in industrial R&D and readily available for hiring by MNCs in Singapore; and
- Developing appropriate capability to perform joint R&D with industry partners and willingness to allow their access to specialised equipment and expertise in the RIs.

In order to attract and root the high-value added manufacturing operations of MNCs in Singapore, it was also critical to ensure that the supporting industries comprising local SMEs developed attractive high-tech indigenous capabilities. Up to the early 2000s, MNCs sought reliable supporting industries among the SMEs and then transferred the needed technologies to them.

---

[9]Chesbrough, Henry. Open Innovation: The New Imperative for Creating and Profiting from Technology. Boston, Harvard Business School Press, 2003.

However, as the pace of technology innovations quickened in the late 1990s, this more leisurely pace of technology transfer also changed. With the MNCs having to speed up the production and marketing of new products to stay relevant in the market place, there was no longer the luxury of time to transfer technology and train SMEs to handle the new technologies. The preference would be to go to a country where they could source their components or processing needs from local SMEs already trained and equipped with the relevant technologies. This trend would gain more momentum as the increasingly global manufacturing practice of outsourcing had resulted in a reduction of in-house manufacturing expertise in many of the MNCs themselves. Hence, the presence of strong local SMEs built on good R&D would be an incentive for MNCs to relocate their home-base-exploiting R&D sites in a country with such facilities. The available R&D manpower, knowledge and other technological structure could even attract foreign SMEs from developed nations to set up subsidiaries in Singapore as they seek to globalise their business operations. A number had in fact been attracted to the Science Park and taken advantage of the tax incentives to establish themselves in the 1980s. Diagnostic Biotechnology Pte Ltd established in Science Park to research, make and distribute diagnostic products worldwide, focusing on immunodiagnostics. Its product range included enzyme immunoassays (ELISAs) and Western Blots for the AIDS viruses, HIV1, HIV2 and HTLV1, and for the dengue fever virus. Another original Science Park I tenant was a company started up by an NUS Biochemistry Department professor, Dr Tan Kok Kheng, who had been doing research on mycology and had incorporated a spin-off to grow mushrooms on wood waste for the local market. Everbloom Mushrooms was incorporated in 1980 and began production in 1983. It was NUS's first commercial spin-off from research.[10] A prominent tenant of Science Park I in 1984 was Seagate, which set up a development facility to support its manufacturing facility in Singapore. Another was a company that made artificial heart valves, Singapore's first such company. Between 1983 and 1987, Science Park took in 34 R&D set-ups.

The need for MNCs to speed up manufacturing of new and innovative product lines to remain relevant in the market place could create opportunities for an open R&D ecosystem like the one Singapore was looking into developing. Such an ecosystem would also be relevant to local SMEs with ambition and the potential to become domestic MNCs themselves. If they had access to research institutes and the research capability of universities in Singapore to

---

[10] Information from Dr Tan Kok Kheng.

meet their technological needs, such SMEs could better realise their potential, in the way that MNCs had been able to grow on the research capabilities in their home countries.

In order to accelerate the development of national research capabilities and capacity to help realise the national plan for a knowledge-based economy, a number of national research institutes in different areas would need to be created to focus on strategic areas of economic growth. Such institutes would attract both local and foreign talent. As one of its national contributions, the research institutes would plan to "flow" researchers regularly to industry so that such talent could contribute more directly to industry. A healthy flow of "human transfer" would thereby be created. Since they would need to maintain a critical mass of experts, research institutes would continue to replenish their talent pool by aggressively recruiting both locally and worldwide. Former NSTB chairman Teo Ming Kian said in an oral history interview in 2014: "In MINDEF we were always talking about being a smart user and being a smart buyer. Why? Because interestingly, smart begets smart. The more people know that you know, the more they are prepared to share and work with you, and the more you are able to build yourself up. So in a way it is a sort of virtuous cycle.... These are, in a way, the key mindsets, the mental frame of mind at that point in time, building capability, trying to be smart in a way, so that we can create a virtuous cycle. And that was the role that I defined for NSTB: bring about creation and application of knowledge and innovation, so that we can sharpen our competitive edge and maintain our relevance to investors to the world." Thus, one key part of the economic restructuring strategy was to develop national R&D capability and capacity. Our research capability hinged on the ability to draw in the R&D capabilities of MNCs, boost the technological strengths of local SMEs and eventually to develop spin-offs, niche services and products to build diversity into the economy to meet changing circumstances and new trends.

Besides getting the research institutes up and running with public money, the first National Technology Plan also strengthened the existing R&D incentive schemes that were aimed at promoting more private sector investment. One existing scheme that was expanded with more funding was the ongoing Research & Development Assistance Scheme (RDAS) (see Chapter 1 for more details). The scheme had succeeded in encouraging existing MNCs as well as local SMEs to engage in R&D projects, either on their own, or in collaboration with other public R&D or tertiary institutions. The scheme encouraged in particular projects that would lead to the enhancement of the company's competitiveness and in-house development capability. One of the local companies that benefitted from RDAS in the 1990s was Keppel-FELS Shipyard which used the grant to develop an offshore jack-up rig of their own design (see Chapter 8 for more details). Another project supported by RDAS was research by a biomed company, Singapore Biotech, which in 1992 began research on a diagnostic assay for the virus which causes a new variant of Hepatitis B. The kit came on the market in 1994. The new diagnostic system means that patients suspected of developing a variant of Hepatitis B only had to wait two hours for a diagnosis instead of two weeks during which the samples had to be sent to overseas labs for analysis. Earlier diagnosis means earlier treatment, a vital factor as the variant illness is more severe. Another company that applied for an RDAS grant was Hyflux which proposed to research the purification of waste water for use in the Singapore Zoological Gardens (see Chapter 8 for more details). Its underlying purpose was actually to research membrane technology. In 1995, 23 projects were approved under RDAS with grants amounting to $13.5 million. The corresponding industry commitments for the same projects amounted to $25.7 million.

In 1992, NSTB introduced Research Incentive Scheme for Companies (RISC) to support companies wanting to invest long-term in R&D programmes or facilities to build up core technological capabilities. Teo Ming Kian said: "We created a scheme called RISC — Research Incentive Scheme for Companies. Then we were debating whether we should call it RISC because first of all, companies may think that they were going into something very risky. We decided to go ahead with it, be upfront and highlight that yes, R&D particularly in a totally new location is risky but we were prepared to come and support and mitigate that risk. So we decided that yes, might as well be upfront and call it RISC. So these I think were important to demonstrate our commitment to create that environment that's conducive for research and development." RISC was designed to attract MNCs and other companies with manufacturing facilities in Singapore or the region to develop their own R&D facilities in Singapore in support of their growth. The scheme proved to be popular. For instance, between 1993 and 1997, the scheme awarded grants to 40 companies.

Teo was quoted in A*STAR's commemorative publication as saying: "I was often referred to as the Billion Dollar Man as the NSTB was allocated $2 billion for its first five-year S&T plan. Even though funding support was often not the most critical factor for a company's decision to locate their R&D centre to Singapore — since it would still have to bear the bulk of R&D expenditure — the scheme signalled the commitment of Singapore to R&D and helped the companies draw in researchers to their laboratories. This scheme was responsible for building up private sector R&D capability quickly." An illustration of resilience was that in the depth of the Asian financial crisis from 1997 to 1999, the RISC programme was still able to attract $1 billion of private sector R&D investments in each of those years. Like RDAS, RISC and other incentive schemes to attract R&D setups in Singapore, the strategic idea behind incentivising R&D was to add value to the manufacturing processes.

The establishment of NSTB was to be the first vital step of the research journey that ushered in a new era and propelled Singapore into the top ranks of research-intensive nations in less than 15 years. By end of 2000, NSTB had implemented two National Science & Technology Plans with a total budget of $6 billion. The promotion of R&D in industries was successful as shown by the proportional increase in industrial R&D spending — keeping the private sector share of the total R&D spending at 62%. The total R&D spending reached a historical high of 1.82% of GDP, while the number of RSEs per 10,000 labour force increased by 2.4 times over 10 years to reach 66, up from the pre-NSTB figure of less than 28 per 10,000 labour force. The number of PhD researchers reached 3,111, a 3.2 fold increase during this period.

The strategy of hosting the NSTB-funded research institutes and centres at the two universities turned out to be win-win for both sides. The progress of the R&D programmes carried out in these RIs and research centres was accelerated as the new research buildings, new equipment and manpower recruitment were facilitated by experienced university administrators. University staff and PhD students had access to the state-of-the-art equipment through R&D collaborations with the RIs and research centres. Job opportunities opened up for PhD graduates in the RIs. University professors were first attracted to participate in the applied R&D agenda; some were more highly motivated to start longer-term use-inspired basic research funded by the Ministry of Education or co-funded by the research institutes. The presence of active NSTB RIs on the campuses created conditions for the universities to become more research-intensive.

NSTB also continued and expanded the scope of the national science and technology awards pioneered by the Science Council to promote science and engineering by recognising outstanding researchers and research leaders during the annual technology month events. The awards include the Young Scientist Award in conjunction with the Singapore National Academy of Sciences, National Science Award, National Technology Award, and National Science & Technology Medal. (See Appendices: Honouring Scientific Talent)

Institutes of Advanced Engineering

The Institutes of Advanced Engineering had been proposed and approved in the 1990 Development Project Committee (DPC) Paper. The goal of the institutes was to jump-start R&D in the areas of economic interest to Singapore.

The approval process for the first two institutes was not smooth in spite of their strategic importance. There were the typical start-up issues that arise when there is a major change of direction and when a relatively big investment is called for. The DPC Paper submitted by Prof CC Hang and Prof Brian Lee ran into "Dr No" in the person of Ngiam Tong Dow, Permanent Secretary of the Ministry of Finance (MOF). The proposal for the two institutes was rejected on the ground that it was cheaper to train R&D manpower by sending Singaporeans overseas to study and work in already established research institutes. The stalemate prompted Prof Hang to turn to EDB chairman Philip Yeo to see if he could do anything. Surprised by the rejection, Yeo immediately sent a note to Ngiam asking for a reconsideration as these Institutes were essential for Singapore's future economic development. Ngiam then sent Deputy Secretary Low Sin Leng to meet up with Yeo and Prof Hang. She explained that MOF had not budgeted for such a large amount of money for research. However, the Ministry could consider giving half of the projected budget — $100 million for the two Institutes. Yeo then asked: "CC, can you do it with half the budget?" While he thought the budget was too low and restrictive as the plan was rather ambitious, Prof Hang did not turn it down, replying: "Half is better than nothing. I shall see how we can manage to get the Institutes started!" It was a "leap of faith" response as Prof Hang, like other academics in Singapore at that time, had no experience at all in establishing large-scale research institutes. After assessing the revised submission, MOF approved a budget of $81.56 million for the two institutes. Some eight years later, when Prof Hang had the rare opportunity to be seated next to Ngiam during a lunch hosted by Hyflux, Ngiam told him that he was pleased with the role of NSTB and the research institutes in helping companies, including local enterprises in developing their R&D capabilities. Puzzled by this comment, Prof Hang then reminded Ngiam that he was the Permanent Secretary who had originally rejected the DPC paper for setting up the two RIs. Ngiam then gave the surprising answer: "Prof Hang, you did not do your homework. In the Ministry of Finance, my nickname is 'Dr No'. When I saw a very large budget request, I would say 'No'. And most of the time, the request went away. But yours came back, indicating that you were serious and willing to fight for it. So I gave it a second chance and tested you further with half the budget!" (In a 2015 conversation between the two men who became friends, Ngiam revealed that he learnt his frugal way of approving the Budget from his first Finance Minister, Dr Goh Keng Swee.)

When it was decided that the first five-year National Technology Plan should be kick-started with the first of two Institutes of Advanced Engineering, namely Institute of Microelectronics (IME) and the Institute of Manufacturing

Technology (IMT), the implementation became the next biggest challenge. At that time, there were no experts in the public and private sectors who could assume the executive director positions in the institutes to lead their rapid establishment and development. Two start-up strategies were conceived and adopted. The first strategy was to locate the institutes at NUS and NTU.

Both universities had established strong engineering education programmes and while research funding was small, there were academics who were doing defence technology research in the late 1980s for MINDEF which had bigger R&D budgets. NUS had already gained some experience in adding an applied R&D agenda to the Institute of Systems Science, and in starting up the Institute of Molecular & Cell Biology (IMCB) which had been officially opened in 1987. NTU had also gained relevant experience in hosting the Grumman International NTI CAD/CAM Centre (GINTIC). Apart from the experience that the universities had gained in hosting these early institutes, other reasons for continuing to have the universities host these two new NSTB institutes were that both universities had sufficient land to accommodate new buildings for them. At the same time, the universities could provide the necessary administrative support and branding. NSTB had been conceived of as a funding agency. Becoming A*STAR and host to the RIs in 2002 was a sea change for NSTB. In 1991 the advantage of having the RIs "piggy-back" on the universities was that with the universities hiring the researchers for the institutes, the researchers had a greater sense of security. In the unlikelihood that the institutes should close, the universities could absorb them. There were other reasons. Teo Ming Kian explained: "The decision to host them in either NUS or NTU was to facilitate collaboration and networking with the existing research community which was primarily in the universities. Another reason was to encourage the universities to increase their impetus to strengthen their research capability."

The second strategy outlined in the first National Technology Plan was to tap into the huge store of excellent ties that had been built up in the past decades with MNCs by public agencies such as the EDB, Science Council as well as the international network established by the university academics.

Attracting International Talent

The EDB's numerous links with CEOs of MNCs all over the world and the excellent reputation it built up for Singapore as a country that delivered in its growing years were a tremendous help in building the R&D landscape between 1991 and 2015. For these were the connections that brought the much needed expertise to start first the training institutes, then the research institutes and to draw in the scientific talent to helm the infant institutes, and the high-powered industry CEOs and CTOs and international researchers to sit on the advisory panels of the institutes and councils. Many were generous with their help and gave time to sit on the International Advisory Boards that were being set up in the developing R&D landscape. One of their major contributions was and remains helping the RIs to identify international RSE talent and research relevance to industry. Networking and being quick to spot opportunities saw the start-up of the Institute of Systems Science (ISS) in collaboration with IBM. A team of IBM engineers from IBM's TJ Watson Research Centre led by Dr Ifay Chang, with Dr Juzar Motiwalla as the local head, started ISS. In 1981, Singapore had put computerisation on the national agenda and the Institute was formed primarily as a teaching institute to help train IT professionals. IBM at that time was the leading mainframe manufacturer and Singapore's national computerisation drive was expected to be backed by such mainframes. The examples of ISS and GINTIC underscore the formidable links that the EDB, through years of courting the MNCs, had built up with their CEOs and CTOs.

When NSTB established the Institute of Microelectronics and the GINTIC Institute of Manufacturing Technology (formed from a merger between GINTIC Institute of Computer Integrated Manufacturing and the Institute of Manufacturing Technology to become GIMT and subsequently Singapore Institute of Manufacturing Technology (SIMTech)), it was very fortunate that two identified international experts were persuaded to serve as the founding executive directors. Dr Bill Chen who founded the Institute of Microelectronics was an AT&T Bell Labs researcher in his mid-50s with years of experience in the laboratory as well as in manufacturing and R&D management. He had earlier served as an international advisor in microelectronics on the Science Council's first International Panel of Advisors (IAP), as well as being one of the international advisors who had helped Taiwan to build up its semiconductor industry. When EDB chairman Philip Yeo asked AT&T for help, Dr Chen indicated his interest. Dr Chen committed to a three-year term but ended up staying for 12 years. Dr Chen recruited 15 senior RSEs, all expatriates with many years' experience in the industry. Many young researchers and fresh graduates were then recruited and trained by these experienced seniors. Prof Hang, founding chairman of IME, acknowledges that without Dr Chen, it would not have been possible to have grown the institute so quickly. (Box: The Institute of Microelectronics)

The setting up of what became the Singapore Institute of Manufacturing Technology was also done through networking. Dutchman Dr Frans Carpay

Institute of Microelectronics: Triggering a Decade of Personal Ventures
Bill YS Chen, Founding Director of Institute of Microelectronics. Abridged from an article written in 2002.
(Dr Chen was awarded the National Science and Technology Medal in 2001)

In the late 1970s, there were many Chinese-American scientists and engineers who served as enthusiastic voluntary consultants in Taiwan's push to develop high technology. Thanks to diligent efforts by the EDB, our interests spilled over to Singapore. In 1981 and in 1983, at the mini-Modern Engineering Technology Seminars in Singapore, I was invited to talk about the design of integrated circuits (ICs). My impression was that acceleration in science and technology was needed. "This is a place where I can lend a special hand when the time comes (as a second career, I meant)." I said this to myself and also incidentally to *The Straits Times* reporters. I was only in my mid-40s then.

EDB and the Science Council subsequently established an International Panel of Advisors on Science and Technology headed by Prof CN Yang, a Nobel Laureate in Physics. Seven of us were on the panel and we were invited to visit universities and some MNCs. We met many interested members of the Government including Messrs. Tony Tan and Lee Hsien Loong. Our impression was that, while the needs of the industries were already pressing, the university graduates did not seem to get that. Our final recommendation was for Singapore to set up institutions to attract students into R&D as well as graduate studies.

By 1990 I had been Director of IC Customer Service Laboratory in AT&T Bell Labs for five years. While on a course in the Catskill Mountains in New York State, I got words from Prof Choo Seok Cheow, Chairman of Singapore Science Council, that Singapore was setting up a new institute focusing on microelectronics. Prof Choo informed me that a Protem Committee had been set up by NUS to put together an initial set of scopes for the institute and that EDB, through its chairman Philip Yeo, had started inquiries with several large MNCs in search of nominees for director candidates for this institute.

In September 1990, I was invited to visit Singapore to explore officially the possibilities. I met the entire Protem Committee headed by Prof CC Hang, as well as Philip Yeo, then Chairman of EDB. I was quite impressed by the firm commitment shown. We soon reached agreement in principle that I would take on the job and on what the scope of the institute might be. My family and I decided that to avoid any possible conflict of interests I should retire rather than take leave of absence from AT&T. However, my boss, Dr Mark Melliar-Smith, VP R&D, AT&T Microelectronics, promised to "un-retire" me, should I decide to go back after three years. It was a rare offer from AT&T. I was then 55, a perfect time for a second career.

(Continued)

(Continued)

An Unlikely Birthplace — PO Box 373 Martinsville, NJ

Highest on the list of my responsibilities as Director of the Institute of Micro-electronics (IME) was to recruit highly experienced senior and managerial staff with rich industrial R&D experiences. I made the US my recruiting base by placing advertisments in the US and UK for senior and managerial talents worldwide. Those interested were asked to mail resumes to: International Advisor, Singapore Science Council, PO Box 373, Martinsville, NJ 08836, USA; Fax 201-722-6027.

As the US economy was weak in early 1991, I was flooded with more than 800 resumes from the US and UK. Out of 80 offered positions, about 30 finally came to IME. Specifically, we were looking for talents in Microelectronic System Applications (MSA), Microelectronic Processing Technologies (MPT), Failure Analysis and Reliability (FAR), and VLSI Design and Test (VLSI). One of the candidates that I travelled across the US to interview was Dr Tan Khen Sang. He was working in Texas Instruments (TI) as a highly regarded Senior Technical Staff on Mixed Signal IC design. He was interested but was too busy working on an important project. Somehow I forced his hand by flying down from Newark to Dallas and caught him for a couple hours after work. I am glad that he eventually did join IME, built a strong world-class mixed signal and RF IC team and eventually took over the helm of the institute eight years later.

By the fall of 1991, we had put together a strong group of R&D Managers (RDMs). They had had years of experience in industrial R&D in the US. Without any doubt, IME owed its early success to them.[11] To this list I must add Prof Chua Soo Jin from the NUS EE Department. A member on the Protem Committee as well as IME's Assistant Director, he acted as my local liaison before I moved to Singapore. Immediately, Prof Chua and the RDMs went to work, especially on the most urgent task of recruiting senior R&D staff to head projects. Our target was to gather at Singapore by the end of 1991. And we did!

---

[11] This is the roster of IME's founding RDMs (in chronological order of their joining IME), the companies from which they came, and their functions in IME in 2002:
- Mike Randall, Intergraph, FAR (returned to US in 1995);
- Dr Victor Huang, AT&T Bell Labs, MSA (later became IME Deputy Director before returning to US in 1996);
- Dr Pang-Dow Foo, AT&T Bell Labs, MPT (Nanyang University graduate, DSIC Program Director and Associate Director in 2002); and
- Dr Tan Khen Sang, Texas Instruments, VLSI (Graduate of Chung Ling High in Penang, Director in 2002).

(Continued)

(*Continued*)

Champions and Friends of IME

The establishment of IME would not have been possible without its champions and friends. First, there were Prof Lim Pin, Vice-Chancellor of NUS, and his deputy, Prof CC Hang. Both endorsed enthusiastically IME's mission to add value to industrial partners through relevant R&D. They were supportive when we decided to establish IME's facilities outside the campus to be closer to the industries. In particular, I am to this day grateful that Prof Hang intervened when IME's clean room budget (about $30 million) ran into a funding problem because of unclear industry directions in early 1992 and placed the entire sub-micron IC technology programme in jeopardy. Prof Hang managed to convince the powers-that-be to let us go ahead with setting up a clean room in Cavendish, Science Park I, where prototypes of integrated MEM devices would be fabricated. Others in Singapore, including industry and university representatives, supported IME by participating in its management board.

EDB in particular gave its sustained support to IME in the promotion of relationship with industries, and in introducing to IME potential International Advisors from MNC headquarters. EDB introduced me to several company executives in Europe and Japan and I visited a number of key laboratories there: IMEC at Leuven, Institute for Microelectronics in Stuttgart, Fraunhofer IMS in Duisburg, and Sumitomo's GaAs facilities in Osaka. Wherever I visited, I was welcomed with open arms. Many of these executives later became our IAP members.

Among all IAP members, Tsuyoshi Kawanishi, previously Toshiba Group VP and subsequently CEO, served IME the longest. He was an IAP member from IME's inception and became IME's Chairman of Board from 1997 to 2001. He helped IME to recruit new IAP members from equipment manufacturing companies in the mid-1990s. He introduced IME to Japanese industry counterparts working on electronic packaging, in-line metrology and 12"-wafer equipment topics. Other IAP members included Dr Horst Fischer, VP Siemens, who advised IME on the practices of various industrial labs in Europe and eventually helped set up visits to these labs, and Dr Pasquale Pistorio, CEO of ST Microelectronics, who familiarised IME with ST's areas of focus and encouraged collaborations. Finally, Dr Mark Melliar-Smith, who was my superior in AT&T at the time of my retirement and was CEO of Sematech until 2001, and who helped IME gain a broader view of the entire semiconductor industry.

From Chai Chee to Pasir Panjang

While we were busy recruiting, visiting industries, and starting to order equipment needed to kick things off, it was clear that we needed a decent place of our

(*Continued*)

(Continued)

own before we could even begin to think about a new building. In January 1992, we moved to 750E Chai Chee Rd, #7-03/04, Chai Chee Industrial Park.

After more than one year working with the industries, it became clear that IME had to expand its activities to include electronic packaging as one of IME's main stakes. In May 1993, we established the APDS (Advanced Packaging Development Support) department under the leadership of Dr Lim Thiam Beng, a Senior Technical Staff from Texas Instruments, (Sg) Pte Ltd. Several local engineers followed him to join IME. APDS spent the early days on current engineering to support production. Gradually, it moved toward forward-looking projects with heavy participation from HQs of the MNCs.

By the end of 1993, IME grew to be 80-member-strong, with five departments and a full range of support in HR and Financing. Industry customers' needs determined our directions. Among the various projects, the contract with Heimann on accelerators together with (mixed) signal-processing ICs and the design of MPEG-2 Decoder were two major ones. We were offered a site in the newly-developed Science Park II, off Pasir Panjang Road. The proposed Information Technology Institute (ITI) and IME were to share a common building which would become Science Park II's first. IME (and ITI) moved into the new building in April 1995.

Shaping TeamIME's Character

Character-building is a necessary journey of aspiring organisations. One important characteristic that we wanted for TeamIME was to build its reliance on collaborative effort and to set up formal ways to measure the effectiveness of collaboration. Another important principle was our ceaseless attention to quality. Last but not least was to maximise opportunities to learn from our counterparts in any collaboration. Chronologically, attention to quality came first. We decided to take

(Continued)

had been with the Philips R&D centre in Eindhoven and who was experienced in the whole R&D cycle from research to commercialisation had been recommended by James Boyd, the founding director of GINTIC which had been established in 1985 and which subseqently became the GINTIC Institute for Computerised Manufacturing. Boyd had been on secondment from Grumman International, the "loan" having come about because the Singapore Armed Forces had bought equipment from Grumman International in the 1970s. Boyd had strong links with the

(Continued)

on the task to have IME certified. It would force us to document the processes based on which we would conduct our business and to stand ready to be audited. By May of 1996, IME was certified with ISO 9001 and a year later ISO 9002. IME has maintained its certified status ever since. Our quality policy, as stated on the back of every IME staff member's name card, has been:

"IME is dedicated to add value to our partners, and to ensure that our R&D processes & procedures are compatible with their manufacturing practices and with the needs of their customers."

The next thing to tackle was to encourage collaborative efforts amongst our staff. We wanted to have clearly defined tasks (business units) participated by several R&D departments whose basic functions were to build specific core competencies. The tasks were to have well planned deliverables and target incomes. That is the essence of the special breed of matrix management practised in IME. The purpose of learning is to ensure that we could serve industry's needs without becoming outdated ourselves. An implantation process by bringing in new blood has a significant shortcoming in that it might not root. Rooting is critical if we want to see our effort grow in future years. We decided to create learning opportunities through our collaboration with customers, which is not easy to do. We learnt about testing smartly tricky MEMS devices from our counterpart, Knowles. We learnt about programming DSP with parallel processors by working with Hitachi on MPEG4. We learnt about RF systems by working with OKI on PHS phone systems. This knowledge after being properly digested by our staff helped us in many other projects.

Looking back, I think we have managed to trigger off a decade of transition in the manner R&D is done. For me personally, it was a decade of personal ventures. I hope that we have left behind us not just a building with sophisticated modern facilities and an institution with dedicated people, but also a way of life, productive, exciting and rewarding in itself.

Dutch R&D network, while Singapore itself had a strong link with the Dutch MNCs courtesy of Dr Albert Winsemius, the Dutch economist who in the 1960s had come to Singapore on a UN mission to advise Singapore on ways to develop the Singapore economy. Dr Winsemius had stayed as economic advisor to the Government till 1984. Dr Carpay had been identified to take charge of a new Institute of Manufacturing (as outlined in the 1990 DPC paper) but found himself merging his new institute with the GINTIC Institute of Computer Integrated Manufacturing to form GINTIC Institute of Manufacturing.

Dr Carpay arrived with his family in Singapore as a secondment from Philips in 1993. Instead of staying the agreed three years, he, too, remained involved with Singapore's R&D evolution from helping to prepare the 1996 and 2001 National Science & Technology Plans to serving on various advisory committees including T-Up. After an involvement of more than 12 years with NSTB/A*STAR, Dr Carpay remained as a visiting professor at NUS, developing and teaching a new course on Management of Industrial R&D. He finally retired in 2014 aged 74.

International expertise is invaluable even in today's digital age of information and connections via the Internet, as there is always confidential or advanced knowledge or even tacit knowledge[12] which is not so easily available. The panels of international advisors which are made up of scientific and engineering talent and industrialists play a role in setting directions and benchmarking research developments. The practice of benchmarking Singapore's science and technology developments with a panel of international advisors had been started in 1985 when the Science Council set up two panels of international advisors, one on physical sciences and engineering, and the other on biological sciences. The most prominent member in the physical sciences and engineering panel was the first Nobel Laureate to be associated with Singapore's S&T development: Dr Chen-Ning Yang who had been awarded the Nobel Prize in Physics in 1957, together with Dr Tsung-Dao Lee.

Another programme that has brought more scientific talent here is the Visiting Investigatorship Programme (VIP) introduced around 2000. A*STAR set up VIP with the aim of attracting eminent international scientists to lead R&D programmes in the RIs, in niche or emerging areas or to catapult our accumulated capabilities in a field to a much higher level in a short span of time. The VIP was a modification of the Temasek Professorship Programme established in the late 1990s by NSTB (See Chapter 6) where leading international scientists conducted their research in the universities and built expertise that currently still resides in various departments and centres. Many core competencies in the RIs today also originated from the VIP such as organic electronics, biomaterials, atomic engineering, molecular diagnostics and computational materials science. The VIP remains an active programme in A*STAR where future capabilities in areas like stapled peptides, cognitive science and glycobiology (to name a few) are being supported.

After 2001, networking and attracting international talent would play an even more critical role in getting the Biomedical Sciences Initiative started. International research talent started up some of the key research institutes that would form the bed-rock of Singapore's Biotechnology Plan. After 1983,

---

[12] See footnote, page 207.

Singapore's most notable scientific advisor has been Dr Sydney Brenner who first came as the keynote speaker of a Science Council seminar on biotechnology and ended up becoming scientific advisor to the Government on biotechnology. This particular economic strategy can be traced back to Dr Goh Keng Swee, to the beginning of the 1980s. (See Chapter 5, Whales and Guppies)

Today, leveraging on Singapore's international scientific and industrial connections, Singapore's R&D ecosystem has impressive advisory panels for the institutes and the Councils comprising men and women of S&T as well as industry. The panels convene regularly — and today through the Internet and digital technologies as when necessary all year round — to look at trends and directions in research. There are some 70 nationalities in Singapore's R&D ecosystem. Increasingly, Singapore's R&D leaders are getting invitations to sit on the international advisory panels outside Singapore, no better indication of how far Singapore's research institutes have come as well as how credible is the R&D being done here.

The Singapore R&D ecosystem is the subject of study by other countries but whether it can be replicated wholesale is another question. For one thing, the strategy of attracting notable international scientific talent to jump-start the research institutes and its multinational advisory panels was built on the links established by the various agencies involved in the economic development of Singapore, including the universities, and the linking began even before independence in 1965 when self-governing Singapore after 1959 first turned to the United Nations Development Programme for help in working out an economic development strategy. By the 1990s, Singapore had the wherewithal to incentivise the movement of top scientific talent with big research budgets, state-of-the-art equipment and various perks to what was essentially an R&D backwater to help turn it into a credible international R&D hub.

By 2000, Singapore had caught up with other high-tech economies like Taiwan in selected areas of microelectronics, IT, disk drive technology, and materials. The reason Singapore was able to achieve this goal against shifting benchmarks (as these advanced economies were also improving) was not only due to its head-start enabled by international talent but also by the fact that Singapore being English-speaking and English being the language of global science, it could turn to the established R&D powerhouses in the US and UK. Said Prof Hang: "We attracted the first generation of very senior people who brought the know-how and the systems and then we attracted Asian talent to come and therefore they became the next layer. When A*STAR was formed, we began developing the third layer, our local core of PhD students. So even as the first generation retired, we were able to continue on our own."

Home-Grown Research Talent

There was a small body of home-grown research talent located in the universities. These academic researchers had small-scale research experience. Up till the formation of NSTB, research grants were in the tens of thousands, hundreds of thousands. A million was considered huge. NSTB could tap the emerging local pioneers to provide scientific leadership for the nascent research entities from Singapore's two universities. Among these would be men like Prof Low Teck Seng who was called on by EDB to start up the Magnetics Technology Centre in NUS Faculty of Engineering in 1992. The Magnetics Technology Centre (MTC) was later expanded to become the Data Storage Institute (DSI). Prof Low was then with the Department of Electrical Engineering in NUS.

The research entities that eventually merged into the Institute for Infocomm Research (I²R) also had a string of leaders drawn from the universities. The founding director of the Information Technology Institute that grew from the R&D Group in the Institute of Systems Science was Lim Swee Say, later to become a politician. The Centre for Wireless Communications (CWC) and the Centre for Signal Processing (CSP) were also helmed by professors from NUS and NTU, respectively, such as Prof Tjhung Tjeng Thiang and Prof Lye Kin Mun from NUS, and Prof Er Meng Hwa and A/Prof Ser Wee from NTU. In 2002, when I²R was formed, Prof Lawrence WC Wong of NUS was identified to head the new institute as its executive director.

Also available was scientific talent that had migrated but could be persuaded to come back to help start up. The early years of IMCB was helmed by Dr Christopher YH Tan, a University of Singapore science graduate who had built a career as a researcher in Canada. Two other University of Singapore graduates who went on to develop a career in research were Dr Louis Lim and Dr Chua Nam Hai. All three men were keen on a career in research but due to a dearth of research opportunities in Singapore, they had established themselves overseas. Dr Louis Lim was with a London research institute. Dr Chua Nam Hai had built his research career at Rockefeller University in New York. The first scientific advisory panel of IMCB consisted of Dr Chris Tan, Dr Lim from University College, London, and Prof Chua Nam Hai from Rockefeller University in New York.

Given the research opportunities, home-grown talent were able to maximise these opportunities. Prof Low Teck Seng, founding director of Data Storage Institute (DSI), said in 2011 on what could go wrong with his task of setting up DSI: "To be able to say where you could have gone wrong is to look at what actually happened and what were the ingredients that allowed you to be successful.

One, of course, is being able to tap on people who know how to set up research institutes, run research programmes, and manage science and technology, to help me craft a strategy and a plan. Without that, at worst you will fail, or, at best, we won't be able to optimise the resources we have, because we'll be wasting time and money going in the wrong directions. So, that's one thing that could have gone wrong. Two, if we had not been able to attract the right people to come to DSI. But they're all interlinked. You have to start it right with a strategic plan that focuses on the right areas and you have to get the right people."

Between 1991 and 2000, NSTB established a total of nine research institutes and four research centres. As these institutes and centres were all incubated in the universities, NSTB could also draw on the universities resources in addition to recruiting foreign talent to lead the institutes.

From NSTB to A*STAR

In August 2000 Philip Yeo became concurrently Co-Chairman of NSTB while remaining Chairman of EDB; Teo Ming Kian remained as Chairman of NSTB while becoming Co-Chairman of EDB. The two men swapped positions in 2001, with Yeo becoming Executive Chairman of NSTB while remaining Co-Chairman of EDB for a transition period, and vice versa for Teo. By 2001, Yeo had been with EDB for more than 15 years, and he took over NSTB with a more aggressive agenda for the Biomedical Sciences Initiative. Biotechnology — pharmaceuticals, chemicals, food processing, agrotechnology — had been one of the emerging growth areas identified in the 1986 Economic Review Committee Report. Yeo wanted to create more industry and research nexus, and to further develop human capital for research, particularly home-grown human capital to anchor the research landscape. One of the first things he did to focus on human capital was to change NSTB's name to Agency for Science, Technology and Research or A*STAR, an acronym that he said would be familiar to every Primary 6 student in Singapore because that was the top grade that they all aimed for. That year he introduced the National Science Scholarships specifically to produce scientific and engineering PhDs to meet the long-term needs of the nation. In 2001 there was a major restructuring of NSTB. A*STAR Graduate Academy (AGA) was created to run the National Science Scholarships and other graduate fellowship programmes that would develop more scientific talent for Singapore. And at the same time, two councils were created — Science and Engineering Research Council (SERC, Chapter 4) and Biomedical Research Council (BMRC, Chapter 5). Exploit Technologies Pte Ltd, a wholly-owned subsidiary of A*STAR, was set up to manage intellectual property and facilitate technology transfer. The restructuring streamlined the national R&D agenda bringing together its research institutes and R&D managers under one virtual roof — A*STAR, rather than scattered across the university landscape. They would all come under two roofs literally when a huge investment in developing physical infrastructure started after 2001. Up till 2001 NSTB had been a funding agency with offices in Science Park. A*STAR was going to be more than that. It was going to take charge of the research institutes literally and to grow the R&D ecosystem into an economic force. A*STAR was also going to embark on growing the Biomedical Sciences Initiative.

The overall achievements of A*STAR in continuing to expand the national research capacity and in ensuring industry's participation are shown in Table 2.1. In spite of the 2008 global recession, industry's R&D expenditure was always higher than that of the public sector. More significantly, the number of RSEs per 10,000 labour force increased to a historical high of 102 in 2010, while the number of PhD researchers reached 7,477, another 2.4 folds increase over the previous 10 years (2000–2010). The overall development of Singapore's economy and the growth trend in industry are illustrated in Figures 2.1 to 2.3.

A*STAR today strives to help Singapore develop into a world-class scientific research hub by building up three types of capital: human, intellectual and industrial. It develops human capital by promoting manpower training and development in the areas of science, engineering and technology (SET). By promoting SET it aims to increase public awareness and understanding of the importance of S&T in Singapore. It enhances and strengthens Singapore's knowledge-creation and innovative capability by directing and undertaking research and development in SET through the research institutes. As the lead agency to promote an economic mission through R&D, it advances the commercial application of scientific knowledge and technology in Singapore through industry engagements and collaborations, R&D investment promotion, and active commercialisation of intellectual property. A*STAR also supports extramural research in the universities, hospitals, research centres, and with other local and international partners.

Table 2.1.   Track record in expansion of research capacity.

Period	1990	1995	2000	2005	2010
Research S$ % GDP	0.85	1.15	1.89	2.36	2.14*
% Private Sector	54	64	62	66	61
Researchers per 10,000 workers	28	46	78	90	102
PhD RSEs	970	1887	3111	4063	7477

* 2008 Global Recession.

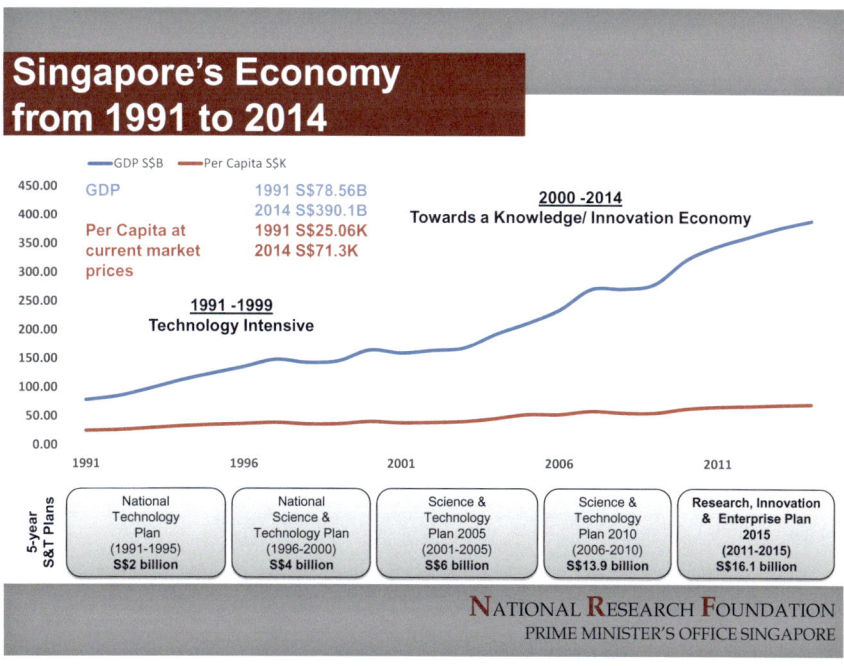

Fig. 2.1. Overall trend in economic development (1990s to 2010s). (Reproduced with permission from National Research Foundation).

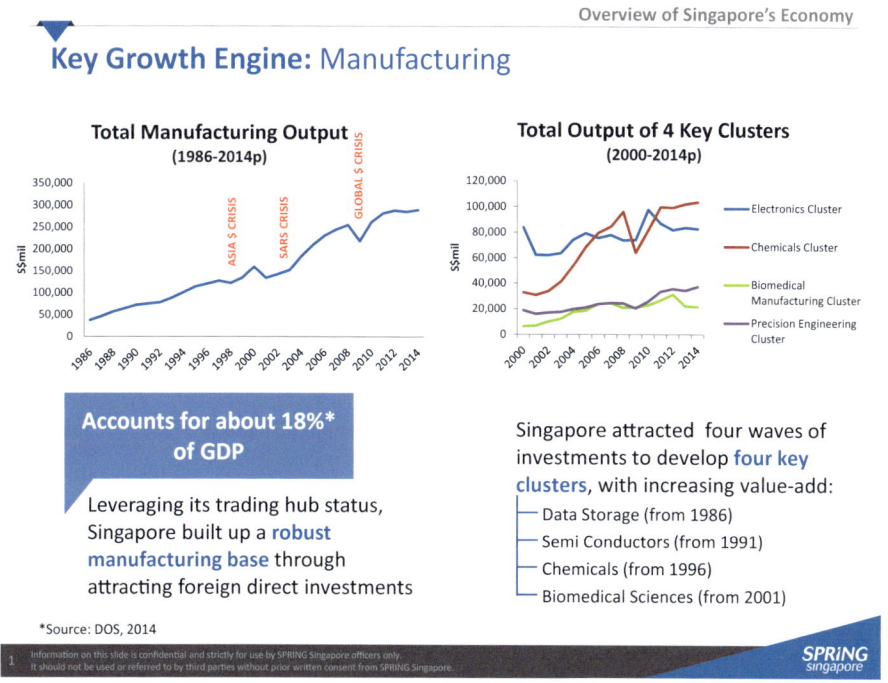

Fig. 2.2. Growth trend in manufacturing. (Reproduced with permission from SPRING Singapore).

Fig. 2.3.   Industry development through the years. (Reproduced with permission from SPRING Singapore).

Chapter 3
The Multi-Agency Approach

Hang Chang Chieh, Low Teck Seng, and Raj Thampuran

Behind the iconic Fusionopolis and Biopolis is a whole complex network of invisible links, local and international, that forms Singapore's R&D ecosystem. It is a multi-layered system that leverages on its links, and whose strength lies in the ties, partnerships and collaborations that span the globe as well as internally within Singapore. These ties have been in the making since Singapore became self-governing in 1959, and economic development, progress and restructuring was adopted as the underlying national philosophy, the benefits of which can be seen today. The principles behind these ties are inherited from the earliest days when Singapore began industrialising to develop the economy and create jobs. In *Heart Work: Stories of How EDB Steered the Singapore Economy from 1961 into the 21st Century*, lead author Chan Chin Bock said in the opening lines of his preface: "Singapore's economic pioneers — former Cabinet Ministers Goh Keng Swee and Hon Sui Sen — have left a very valuable legacy to the Singapore Economic Development Board. It is the legacy of first learning from others, and then adapting new knowledge to apply to the Singapore we want." This is one of the underlying principles that shifted Singapore towards a knowledge-based economy and the development of a vibrant R&D ecosystem. Singapore learns from the more knowledgeable around the world but it also learns from its own history.

The first layer of the R&D ecosystem is the linking of the individual institutes within A*STAR, first through the Science and Engineering Research Council (SERC), Biomedical Research Council (BMRC), and after 2007 the Joint Council Office. The different institutes and various disciplines come together to work on the latest scientific and engineering challenges, and through a multidisciplinary approach is able to advance the frontiers of technology, create innovations, and advance knowledge. Such internal collaborations and interactions help to challenge some of the most difficult technological problems that inherently require contributions from scientists from an array of disciplines. In the 21st century, the tightening of security measures in practically all public and private organisations mandates that such informal interchanges that are used to build up synergy between researchers must now be organised through platforms such as the Joint Council Office, annual fora, conferences or workshops or coordinated collaboration programmes. Another layer of integration is that between researchers and institutes in their partnerships with industry partners be they multinationals companies (MNCs), larger local enterprises (LLEs) or larger small-medium enterprises (SMEs). Such public-private collaborations form another layer of integration enriching an ecosystem of different private-private collaborations.

A very important layer of integration in the R&D ecosystem is public-to-public collaborations. There are several aspects to this. One is the more general collaboration and cooperation between government agencies, ministries and government departments to achieve certain objectives such as facilitating the establishment of a multinational company here. It is about a national effort in fostering partnerships and collaborations. Such collaborations could be in the form of clearance of permits, installation of buildings or manpower support. Collaboration can also be in the form of the administration of certain programmes. For example, A*STAR research institutes (RIs) work with SPRING Singapore which manages the technology upgrading of the SMEs. The second albeit more narrow aspect of public-to-public collaboration is in specific research collaborations between RIs and government agencies, for example, a project between RIs and the National Environmental Agency on studying how dengue fever is spread and its control in the context of Singapore's urban environment. Another is the partnership of RIs with the Urban Redevelopment Authorities on land and city planning.

Public-to-public collaborations can also be at an international level, between governments of different countries, between the agencies in these countries that have a reputation for excellence in science and technology, and between universities. Building ties is something that Singapore has become good at after 50 years of diplomacy and cooperation across borders. While such collaborations are often advisory in nature, they nonetheless help A*STAR, the RIs, universities and polytechnics to set benchmarks from these international and corporate level links.

Public-to-public collaboration and cooperation at the more general national level began with the Economic Development Board (EDB) which was a one-stop agency in its early years (See Chapter 1). Even though many agencies have been spawned from this one agency, EDB continues to have the ability to marshal public sector agencies to support its mission of economic development. A*STAR chairman Lim Chuan Poh has noted how this is a big marketing advantage for Singapore ventures as a whole and not least in the promotion of R&D. He said in an interview in 2011: "Why did Procter & Gamble come here? It's the value creation that we have. When I met the Chief Technology Officer of Procter & Gamble, he was so happy when I told him 'One agreement. You don't have to sign an agreement with 14 institutes. One agreement.' And not only that. I could set up a steering committee with him to resolve any issues [during construction and setting up]. He does not have to go and knock on doors and chase people all around town just to set themselves here. Just that process alone creates value for the investor." It is what Chairman Lim calls "integrating for impact", one of A*STAR's strategies to promote Singapore's R&D ecosystem.

One layer of integration is the building of meaningful collaborations with private sector where the impact of R&D is most pronounced and because economic impact is the key consideration for most public policies. Said Dr Raj Thampuran, Managing Director of A*STAR, in 2014: "It is really always an obsession about what is our value proposition and it goes down to the second level of trying to define what our differentiating attributes or elements are in every single thing that we do. All the time. Every time. Everyone in the Ministry of Trade and Industry and this includes A*STAR, wonders how to grow the GDP and create good jobs. How can I do something to stimulate S&T innovation and raise the value added in our economy? So much is after all predicated on economic progress. Then one can enjoy social benefits, more lifestyle choices, more job options for people and so on."

National Research Foundation (NRF)

An important form of collaboration is the big one between public and private sectors to address national issues. Here, an agency to coordinate and integrate activities and fund projects is essential. Such an agency is the National Research Foundation (NRF) set up on 1 January 2006. One of the national issues managed successfully in the last 20 years through partnerships with both private and public sectors has been meeting the water challenge when supply and reliability came to be highlighted in the 1990s. NRF is a department within the Prime Minister's Office. It is the secretariat to the Research, Innovation and Enterprise Council (RIEC) which is chaired by the Prime Minister to set the overall strategic direction for national R&D and stimulation of enterprise which has been identified as an important engine of growth. The chairman of the NRF Board is the Deputy Prime Minister. This high-level Council underscores the political commitment and importance placed on the national R&D agenda. The RIEC advises the Cabinet on national research and innovation policies and strategies to drive the transformation of Singapore into a knowledge-based economy with strong capabilities in S&T. The ambition is to make Singapore a "Smart Nation" of citizens who maximise S&T applications as *de facto* choices. The Council also leads the national drive to promote research, innovation and enterprise by encouraging new initiatives for knowledge creation in S&T, and to catalyse new areas of economic growth.

The Government has continued to invest further in R&D especially upstream. Unlike A*STAR which focuses on mission-oriented R&D by working closely with industry and EDB, NRF focuses on national issues in its research strategy and planning. NRF focuses on three areas. These are:

- Growing future capability through various initiatives that help develop a strong science and talent base in Singapore. These initiatives include the Research Centres of Excellence (RCEs); Campus for Research Excellence and Technological Enterprise (CREATE); Competitive Research Programme (CRP) Funding Scheme; and NRF Fellowship;
- Supporting future growth through two key initiatives that focus on research with economic and industry relevance: Strategic Research Programmes, and National Framework for Innovation and Enterprise; coordinating the research agenda of different agencies to transform Singapore into a knowledge-intensive, innovative and entrepreneurial economy; and
- Meeting future challenges through a programme that identifies national innovation challenges and funding research to address national needs. (More on the Research Centres of Excellence in Chapter 6)

The Campus for Research Excellence and Technological Enterprise (CREATE) is an international collaboration that houses research centres set up by top foreign universities together with our local universities and research institutes. At CREATE, researchers from diverse disciplines and backgrounds work closely together to perform cutting-edge research in strategic areas of interest, for translation into practical applications that can lead to positive economic and societal outcomes for Singapore. The research centres at CREATE focus on four inter-disciplinary thematic areas of research, namely human systems, energy systems, environmental systems and urban systems. The universities with research centres in CREATE that collaborate with Singapore investigators are: Massachusetts Institute of Technology (MIT); Swiss Federal Institute of Technology (ETH), Zurich; Technical University of

Munich; Technion-Israel Institute of Technology; Hebrew University of Jerusalem; Ben-Gurion University; University of California, Berkeley; Peking University; Shanghai Jiao Tong University; Cambridge University.

The NRF Competitive Research Programme (CRP) Funding Scheme seeks to foster the formation of multidisciplinary teams to conduct cutting-edge research that are of relevance to Singapore in particular and society in general. This allows a coordinated, integrated and sustained way of bringing together complementary research groups in Singapore to conduct high-impact research. The CRP scheme funds use-inspired basic research based on scientific excellence, selected through a scientific merit review process. The theme of the proposed research programme should be motivated by an important need or problem to be solved. Since the inception of the scheme in 2007, NRF has launched 14 calls and awarded 64 projects. Some of the CRP projects have made discoveries of potential significant impact on industry and society.

Under the Strategic Research Programmes, three strategic sectors have been identified:

- Biomedical Sciences Translational and Clinical Research focusing on bench-to-bedside translation of basic biomedical discoveries into better medicines and therapies for improved healthcare. Multi-institutional flagship programmes have been established in gastric cancer, eye disease, infectious diseases, metabolic diseases, and schizophrenia;
- Environmental and Water Technologies including Clean Energy — leveraging on Singapore's foundation in water technologies and management and its competitive advantages in other environmental technology sectors. This is complemented by the Clean Energy Initiative which will focus on solar and fuel cells, in which Singapore also has a competitive advantage. Being a compact city-state, Singapore is also an ideal test-bed for new technologies in these areas; and
- Interactive and Digital Media (IDM) — building on Singapore's multi-cultural, multi-lingual identity and its strong Information and Communications Technology (ICT) infrastructure to create new innovative niches in IDM including animation, games and special effects, education and edutainment, "On-the-Move" media services and media intermediary services.

New strategic areas that have recently been added include Marine and Offshore, Satellite and Space, and Cybersecurity. Attention will also be given to commercial development, bringing scientific knowledge from the lab to the market.

Economic Development Board (EDB)

The primary driver of R&D in Singapore has been the EDB that was set up as a one-stop statutory board to draw in investments, set up industries and basically drive economic development. The lead government agency for planning and executing strategies to enhance Singapore's position as a global hub for various economy drivers from manufacturing to services has shaped Singapore's S&T landscape over the years. The EDB's mission saw its officers being exposed to science and technology applications in industrial settings and realising that Singapore needed to move up the technology ladder if the country were to manufacture for the world market, Thus began EDB's setting up some of the elements of what would form today's R&D ecosystem such as the Industrial Research Unit which would morph into the Singapore Institute of Standards and Industrial Research, subsequently into SPRING Singapore; the technical training institutes that also transformed when the needs became different. Because of the demand for more technically trained manpower, such training institutes were set up in partnership with MNCs, with international partners, and science and engineering education expanded to meet the increasingly complex S&T landscape.

The EDB's ability to draw in technologically more advanced investments meant that the S&T landscape had to keep pace — and also capitalise on it. EDB drew in Beechams Pharmaceuticals in 1972, which set up a plant to produce semi-synthetic penicillin from imported penicillin nucleus. By 1976, Beechams had put in equipment to produce the penicillin nucleus itself, the first in the world to use the latest technology for the manufacture of the penicillin nucleus. Beechams' presence in Singapore was an added incentive for Glaxo to start its plant in 1979 and the presence of these two pharmaceutical companies after 1980 would focus national interest in biotechnology.

The EDB played an essential role in the initiation of the first few research institutes and the establishment of NSTB in 1991. Prof CC Hang, then founding Deputy Chairman of NSTB, recalled that then EDB chairman Philip Yeo would often say "EDB is the father of the Research Institutes, while NSTB is the mother responsible for nurturing the Research Institutes to serve the country!"

A small country like Singapore with limited resources has to be focused in its research efforts, and Singapore's focus is on economic development. Spotting what is economically relevant is not easy and there is no science for choosing the areas on which to invest in research to grow industries. While panels of international advisers all contribute their insights and wisdom, in the end the R&D managers and research directors have to make the final decision. During the early 1990s, the EDB proactively identified the S&T research areas in close consultation with the MNC partners and guided NSTB to invest its resources in appropriate areas. This was an application of the well-tested principle of learning from those with more experience, a strategy that has become part of Singapore's strength and nowhere more so than in the development of the R&D ecosystem in Singapore. The EDB worked closely with NSTB and supported the rapid development of the public research institutes. The availability of research infrastructure and manpower could serve as a supplementary R&D facility or even primary R&D facility for the MNCs, and further down the line the larger SMEs. The presence of such R&D facilities also sustains and rejuvenates manufacturing in Singapore. It is an added value proposition to attract emerging MNCs to consider Singapore as a manufacturing base. So, while NSTB (and subsequently A*STAR) focused on research, EDB continues to help industry players tap into the R&D ecosystem and to develop research capabilities through various incentive schemes.

The Universities

Important layers in the R&D ecosystem have been the universities which in the 1980s comprised the National University of Singapore (NUS) of which Nanyang Technological Institute (NTI) was a part until NTI received its university charter in 1991 to become Nanyang Technological University (NTU). For the universities at that time, producing Singapore's intellectual capital was the focus and thus the priority was teaching and research took a back seat. However, the natural curiosity of academics being such, there had always been some research going on. Dr Gloria Chan who joined the Botany Department of the then University of Singapore in 1957 and who later became head of department and Dean of Science said in *Beyond Degrees: The Making of the National University of Singapore*: "When I joined, research funding was nil. Honestly, there was no money for research at all. So we did what little we could in our spare time and with whatever we could get from the teaching vote. Research was carried out, but on a smaller scale, using less sophisticated equipment." The development of scientific and engineering talent into PhDs and research talent was very limited and would-be Research Scientists and Engineers (RSEs) essentially remained overseas to develop their research careers. A few returned after getting their PhDs to combine teaching with some research on the side. One Singaporean who returned to teach, Dr Tan Kok Kheng from the Biochemistry Department, managed to keep up his research into mycology with funding of about $20,000 from the International Foundation for Science in Stockholm. Dr Tan spun off his findings into a company to grow mushrooms for the local market. It was Dr Tan who presented Dr Sydney Brenner with mushrooms when he visited Singapore in 1985 to inaugurate the Lee Kuan Yew Distinguished Visitor Programme with four lectures on biotechnology. Enough individual research went on in the 1980s for NUS to develop international recognition in certain areas, e.g. Dr Arif Bongso's stem cell research. There was Prof SS Ratnam who was internationally known for his research in human reproduction. Asia's first test-tube baby, a boy, was born in 1983. In 1988, the Department of Obstetrics and Gynaecology of NUS Faculty of Medicine carried out the world's[1] first successful micro-injection pregnancy. The technique called Micro-Insemination Sperm Transfer (MIST) enabled men with low sperm count to father their own children. In 1992, the same department chalked up another first by achieving a human pregnancy from the direct injection of the sperm into the cytoplasm of the egg.

The research scenario in the university had begun changing with the appointment of Dr Tony Tan as Vice-Chancellor of NUS (1980–82). Dr Tan, a doctoral graduate of the University of Adelaide, said in *Beyond Degrees*: "To function properly a university needs to do research. Teaching and research are sometimes regarded as two separate functions. To my mind they are complementary. Whether it is a doctor patiently accumulating clinical observations over many years, a lawyer editing a series of law cases, a historian trying to unravel some puzzling aspects of the nation's past or an engineer trying to resolve the practical problems of industry, a university teacher cannot be a good teacher unless his horizon extends beyond the boundaries of an unchanging set of textbooks and lecture notes." When Dr Tan became Vice-Chancellor (VC), he stated that the university would set aside at least 5% of its operating budget for research.

This focus on research was further developed by Prof Lim Pin when he became VC (1981–2000) after Dr Tony Tan. Prof Lim who was from the Medical Faculty helmed NUS during the critical early years when Singapore was developing its early R&D capabilities. Prof Lim sat on the EDB board and also served as Deputy Chairman of the EDB. As VC, he had set up the Office of Research in 1994 with Prof CC Hang in charge. Prof Hang had by then served the first three-year stint as part-time founding Deputy Chairman of NSTB. In the 1980s more money for university research had become available through the introduction of the Research Development Assistance Scheme (RDAS) in 1981. RDAS grants were larger in quantum compared with the academic

---

[1] The Straits Times, May 29, 1983, p. 20.

research grants from the Ministry of Education, either by themselves or in partnership with industry. The research topics ranged widely from communications engineering to biomedical sciences. Many PhD students and post-doctoral fellows were also recruited from abroad as they could be funded under the RDAS grants. Over the next 10 years, the reputation of university research was given a big boost.

Besides incubating the beginnings of scientific and engineering talent, NUS also incubated the early Public Research Institutes (PRIs). The Institute of Molecular and Cell Biology was a PRI within NUS and was located on the Kent Ridge campus originally when it opened in 1987. I^2R (pronounced I squared R) began life as KRDL or Kent Ridge Digital Labs, as did Institute of Systems Science. Data Storage Institute (DSI) was started as Magnetics Technology Centre by Prof Low Teck Seng. He credits Philip Yeo and the EDB for being the driving force behind the institute's founding: "I'm a university professor. If there wasn't this support for me from outside campus, on what basis would I be able to establish a programme and plan to drive in that direction? Having got that impetus, I was able to articulate a vision and hence, together we were able to build DSI."

Institute of Chemical and Engineering Sciences (ICES) started as a proposal by Prof Hang, then Director of Research at NUS, to set up a Process Analysis and Optimisation Enterprise (PAOE). The multidisciplinary group led by Prof Ching Chi Bun of the Chemical Engineering Department was set up in 1994 with the goal of marrying NUS research capability with industry consultation. The $15,000 grant from NUS for its start-up and planning was subsequently augmented by a $722,000 grant from NSTB as well as another grant from NUS to set up and run PAOE's Industrial Consortium. While NSTB's grant went towards three industry-linked research projects, NUS gave the group additional grants to focus on upstream research that had relevance to the industry-linked projects. PAOE grew in tandem with EDB's development of the process chemistry engineering cluster, with more grants and industry projects coming its way. In 1998, the group was given a $9 million NSTB grant, Prof Ching became its full-time Director, and the name, PAOE, was changed to Chemical and Process Engineering Centre (CPEC) in the Department of Chemical Engineering. It had its facilities in a basement in the NUS Engineering Faculty. In 2000, when Prof Hang invited the NSTB and EDB chairmen to visit CPEC to assess its progress and future plans, EDB chairman Philip Yeo made a surprising suggestion that it should become the research anchor at the petrochemical industry complex on Jurong Island. At the official opening of Jurong Island in October 2000, then Prime Minister

Goh Chok Tong announced the setting up of a research institute in chemical engineering to be called the Chemical Engineering Research Institute. This name then became Institute of Chemical Sciences before it became the Institute of Chemical and Engineering Sciences (ICES). Dr Keith Carpenter who spent much of his career in ICI (UK) was recruited in 2001 to be its founding executive director and ICES moved to Jurong Island in 2004.

Incubation of the nascent public research institutes (PRIs) in the universities came naturally as such PRIs are also incubators for PhD students. With Singapore's R&D landscape being so new, the university connection made for stability and security. It reduced the risk for foreign research talent thinking of uprooting to take up a position in Singapore. As adjunct professors, their role of imparting knowledge was also part of the arrangement with the universities. The establishment of the PRIs enlarged research opportunities for the universities and attracted international attention. Both local and foreign students who wanted to do PhD studies were attracted by the state-of-the-art facilities and generous scholarships. These students could also undertake internships in the NSTB research institutes and they would have access to the research labs if they were jointly supervised by the academic staff and PRI researchers.

The universities as incubators for the NSTB RIs ended in 2001 when the PRIs were reorganised and put under the direct charge of A*STAR. The PRIs, in their national role of building a pool of young, capable researchers as a future supply for industry, provide immediate job opportunities to retain the PhD graduates in Singapore. The two universities have also taken steps to ensure that there are sufficient numbers of undergraduates being inspired to pursue PhD research in due course to meet the long-term manpower needs of Singapore's knowledge-based advanced economy. These include an undergraduate pathway which is more research-oriented, undergraduate research opportunity programmes, undergraduate internship at A*STAR research institutes, student exchange with top overseas research-intensive universities, and final year projects being jointly supervised by academics and PRI researchers.

After the reorganisation of NSTB into A*STAR, new types of targeted public sector research funding schemes were introduced to replace the 1981 RDAS funding. A*STAR would do its strategic analysis, consult the international advisors and the local research authorities before issuing calls to solicit project proposals which would be funded by A*STAR. Such project calls present huge opportunities for university academic staff to access public sector funding and accelerate the development of their world-class labs. The new grant schemes include Science & Engineering Research Council (SERC) public sector funding,

SERC thematic strategic research, SERC biomedical engineering research, Biomedical Research Council (BMRC) general consortia grant, joint-council project grant, etc. There are also new types of competitive research grant calls such as the Singapore-China joint grant call, Singapore-India joint grant call, Singapore-Korea joint grant call etc, which are administered by A*STAR, and university and PRI staff are encouraged to form partnerships and collaborations to submit joint proposals. (More on university research in Chapter 6)

National Medical Research Council

Established in 1994 within the Ministry of Health to oversee the development of medical research by providing research funds to healthcare institutions and awarding competitive research funds for individual projects, its executive role includes taking directions from the Ministry of Health and Biomedical Sciences Research Council advisors in the development of a national clinical research strategy as part of the national BMS initiative, including the identification and prioritisation of areas for clinical research. The National Medical Research Council which is based in Biopolis also coordinates and facilitates the efficient use of research facilities, manpower and funds. In 2006, then Minister for Health Khaw Boon Wan announced: "MOH has updated its portfolio and formally incorporated clinical research as part of its mandate. It is a significant step forward for the Ministry. We will continue to emphasise clinical services, but we will begin to devote new resources to support clinical research. Our focus remains our patients, and clinical research must help enhance our care for our patients." Translational medicine and the building up of a pool of clinician-scientists had become critical to the next phase of the development of the Biomedical Sciences Initiative. It became the responsibility of the Council to develop clinician-scientists through awards and fellowships. Prof Edward Holmes[2] who had been recruited to head the the Translational and Clinical Sciences Group in A*STAR was then seconded to chair the Council to help set directions in this new area. Today NMRC-funded research has led to inter-disciplinary partnerships and international collaborations. It also evaluates the outcomes of the research projects and facilitates the commercialisation of research findings. Since its inception, it has built up the medical research capabilities in Singapore through the funding of more than 1,100 individual research projects and 13 national research programmes. The Council's Board

---

[2]In 2011 Prof Edward Holmes was awarded honorary Singapore citizen for his contribution to developing translational and clinical research.

consists of representatives from the universities and leading medical and scientific institutions in Singapore.

DSO National Laboratories

DSO National Laboratories had its beginnings in the visionary mind of the late Dr Goh Keng Swee. In 1972, Dr Goh, then Minister for Defence, handpicked three engineers to form a secretive research group called the Electronics Test Centre (ETC). Its task was to build up secret-edge R&D capabilities for Singapore in key areas such as electronic warfare, guided systems and cryptography. With the progressive expansion of its R&D scope, ETC was renamed Defence Science Organisation in 1977. In 1997, it was incorporated as a not-for-profit company and became known as DSO National Laboratories (DSO). Today, DSO has grown into an organisation some 1,800-strong and its mission has remained unchanged: to develop secret-edge technologies and solutions that sharpen the cutting edge of Singapore's national security. Many of the Singapore Armed Forces' mission-critical systems bear the invisible imprint of DSO's work.

DSO's research fuses multidisciplinary expertise into leading-edge technologies and systems that cannot be commercially procured. Its seven R&D Divisions undertake a full range of activities that span applied research to exploratory development and full scale development. Key research areas include autonomous unmanned systems and manned-unmanned system teaming; electronic systems including antenna design and electromagnetics; information system including computer security, data fusion, operations research and human factors engineering; networks including *ad hoc* mobile networks, software-defined radios, secured communications and datalinks; sensors including radar, acoustics, electro-optics and underwater; and defence medical and environmental research for protection against chemical and biological threats, as well as the enhancement of soldier performance. DSO also leverages on its R&D capabilities to provide technical evaluation support to Defence Science and Technology Agency (DSTA) and MINDEF acquisition programmes.

DSO's contribution to the national R&D strategy goes beyond sharpening the cutting edge for Singapore's national security. It has also established partnerships with universities, research institutes and industry to work on key research areas. Locally, DSO works with the universities and Temasek Laboratories in NUS, NTU and Singapore University of Technology and Design (SUTD) to conduct long-term research in areas with potential applications for defence. DSO has also established research facilities in the universities. These include the

Singapore Wind Tunnel Facility in NUS, and the Electromagnetic Effects Research Laboratory (EMERL) in NTU. Internationally, DSO is part of the Supèlec ONERA NUS DSO Research Alliance (SONDRA). SONDRA conducts research in advanced electromagnetics and radar, and hosts senior scientists' exchanges between France and Singapore.

Beyond research partnerships, DSO collaborates with industry on advanced development and full-scale development projects. The Advanced Technology Research Centre (ATREC) — a joint venture between DSO and Singapore Technologies (ST) Kinetics — was established to conduct R&D in advanced materials for both defence and commercial applications. Another joint venture between ST Engineering, DSO and NTU — ST Electronics (Satellite Systems) — was formed to design, develop and produce advanced earth observation satellites.

These partnerships and collaborations are a key strategy in fueling science and engineering R&D, as well as in overcoming DSO's limited resources. It also contributes to an integrated technological resource for Singapore and the buildup of critical in-country defence R&D capabilities.

Standards, Productivity and Innovation Board (SPRING Singapore)

The Standards, Productivity and Innovation Board (SPRING Singapore) is an agency under the Ministry of Trade and Industry. One of its tasks is to build a strong base of innovation-centric small and medium enterprises (SMEs) in the economy. The agency aims to enhance the competitiveness and viability of SMEs through innovation and making use of knowledge resources. Thus, it works in partnership with A*STAR to direct such resources to the SMEs that are ready for such technology transfers.

SPRING was previously known as the Productivity and Standards Board (PSB), which was in turn the result of the merger of the National Productivity Board (NPB) and the Singapore Institute of Standards and Industrial Research (SISIR) in April 1996. The goal of the merger then was to bring together the soft skills of productivity promotion that was being handled by NPB and the technical aspects of productivity enhancement as handled by SISIR. In April 2002, it was renamed SPRING Singapore to signify the shift towards an innovation-driven economy, and its new role in promoting creativity to sustain growth for Singapore. Today, SPRING is responsible for helping Singapore enterprises grow and continuing to build trust in Singapore products and services. As the enterprise development agency, SPRING works with partners to help enterprises in financing, capability and management development, technology and innovation, and access to markets. As the national standards and accreditation body,

SPRING develops and promotes an internationally-recognised standards and quality assurance infrastructure. SPRING also oversees the safety of general consumer goods in Singapore.

As discussed earlier in Chapters 1 and 2, SISIR played an important though modest role of promoting the local engineering development capabilities in the 1980s. It continued to play this role in the early 1990s while the NSTB research institutes were busy with building up large-scale R&D facilities before extending their assistance to local SMEs. After 1996 when SISIR became PSB, the agency offered an extensive range of services and incentives to help companies improve productivity by exploiting technology for the development, improvement and commercialisation of products. With the inclusion of its enterprise development role, SPRING has further developed programmes to build up the local innovation and entrepreneurship landscape and to help local enterprises grow through innovation. (More on the role of SPRING in Chapters 8 and 9).

Intellectual Property Office of Singapore (IPOS)

One of the important layers in the R&D ecosystem developed in the last 10 years has been in intellectual property (IP), recognised as a valuable asset in today's globalised knowledge-based and innovation-driven economies. IP is an important driver of economic growth as well as being a transactable asset in itself. The focus on R&D as part of economic restructuring after 1986 has also necessitated an overhaul of the IP laws although awareness of the importance of protecting IP was realised very early on with the arrival of MNCs when Singapore began climbing up the S&T ladder with their help.

In 1982, the Science Council set up a study group to look into the upgrading of Singapore's patent system, inviting a four-member World Intellectual Property Organisation (WIPO) Technical Mission to Singapore for consultations. Despite the early Science Council interest in IP, it was only after the formation of NSTB that ramped up investments in R&D also meant a more urgent need to overhaul the intellectual property laws of Singapore. However, patent protection is not a simple matter of revamping the IP laws. A whole new profession had to be created — the patent profession. Developing IPOS had to go hand in hand with developing the patent profession, one that did not exist in the 1980s. A pioneer in this field is Suresh Sachi who as a young lawyer in private practice handled the 1990s research contracts of the universities and research institutes funded by A*STAR (then NSTB). Today the Deputy Managing Director and General Counsel of A*STAR, he recalled in an oral history interview in 2014: "I remember that at the time we started, contracts were very simple. We were quite unsophisticated. We had to borrow widely from other jurisdictions that were more advanced in this area of licensing and technology transfer. We took reference primarily to what the English and Americans were doing and relied on precedent books mostly from the UK. Over time, we started developing our own expertise and our own tools. Today, I can safely say that we have much more expertise in the area of licensing and technology transfer deal making and our lawyers and technology transfer managers at A*STAR and other public sector research organisations have sufficient skills and the nous to hold our own against top MNCs and their large legal teams."

Singapore in the 1980s did not have its own patent legislation. What it had was the Registration of UK Patents Act and anyone seeking patent protection had to use this piece of law. Singapore then had a Registry of Trade Marks and Patents. The bulk of its work in the early days was the registration of trade marks with a small amount of patents and design work. All such patents and designs applications for protection went through the UK and whatever were registered there were then re-registered here. The Patents Act (original enactment Act 21 of 1994) introduced in 1994 was revised in 2005: "An Act to establish a new law of patents, to enable Singapore to give effect to certain international conventions on patents, and for matters connected therewith." There was a need to develop skills in patent drafting, a highly skilled profession that requires deep knowledge of science and technology as well as patent agent training and expertise in patent law. Prior to the development of the know-how in Singapore, all patent work was done through patent lawyers either in the US or UK and then Australia, the more cost-efficient option.

The 1994 introduction of the Patent Act was a challenge for the legal profession. It became necessary to develop the patent agent profession from the ground up. The newly-set-up Patent Registry in IPOS conducted a patent drafting course run by the World Intellectual Property Organisation and Suresh found himself among the 15 or so lawyers who attended the course. Most of the lawyers who attended the course subsequently became patent agents by virtue of the "grandfathering" provisions for practitioners who had been practising patent law before the Act came into force. "This was an important milestone for Singapore. The promulgation of our own Patents Act meant that we were serious about building a knowledge-based economy and one where our local enterprise and foreign countries alike could gain protection for the fruits of their intellectual endeavours. It also meant that Singapore could be an attractive location for companies to base not just their R&D activities but also their IP and enjoy full protection of the law in a nation that respects the intellectual property of others. At about the same time, the Institute of Molecular and Cell Biology (IMCB) engaged Dr Yosef Kimhi, a technology transfer specialist, who came from the Weizmann Institute and the Yeda in Israel. Dr Kimhi, said Suresh, was a highly experienced technology transfer specialist. Working closely with Dr Kimhi, Suresh took the opportunity to learn more on IP licensing and technology transfer. Over time, he was able to hone his skills in drafting and negotiating IP and licensing agreements on the use and licensing of intellectual property as tradeable assets and part of the ability to generate economic impact from the research undertaken by the various research performers. Said Suresh in 2014: "I was fortunate to be able to learn from an expert who also took the trouble to teach a young lawyer some science and the art of negotiation. Some of the things I learnt from Dr Kimhi I still use today. I then try to pass on this know-how and skills to my younger colleagues. As we learn, we teach and use our experience to help grow the expertise around us and strengthen the ecosystem. Today we are in a much better position than where we were back early '90s where there really was not that much experienced IP and technology transfer expertise in Singapore. We have strong IP laws and a sophisticated Intellectual Property Office with our home-grown capability to undertake search and examination of patents — something that seemed so far away when we first enacted our local patent laws. This was a result of dedicated work and strong support from the Government that understood the importance of IP as business and legal tool and an essential ingredient of an innovation economy."

The Intellectual Property Office of Singapore (IPOS) is a statutory board under the Ministry of Law. IPOS advises and administers the Intellectual Property (IP) regime, promotes its usage and builds expertise to facilitate the development of Singapore's IP ecosystem. With IP fast becoming a critical asset in today's global markets, IPOS aims to be a trusted partner to empower all creators in Singapore's knowledge economy through greater recognition of their innovations and to globalise our knowledge economy with a connected and inter-operable IP ecosystem. International surveys consistently rank Singapore's IP regime as one of the best in the world. IPOS's vision is for Singapore to be an IP Hub of Asia. According to the World Economic Forum, Singapore is ranked second in the world and the best in Asia for intellectual property protection in a 2014/2015 report.[3]

To fulfil this vision, IPOS reaches out to various stakeholders:

- For businesses, IPOS continues to provide tools and information to enable them to create, own, protect and profit from their ideas and knowledge;

---

[3] Long, Susan, ed. Things to Love About Singapore. Singapore, The Straits Times, 2015. p. 153.

- For IP professionals, IPOS seeks to upgrade their technical know-how and expertise, as well as provide opportunities for IP professionals to network and exchange views with IP thought-leaders around the world;
- For international stakeholders, IPOS strives to further its cross-border IP cooperation so as to provide a strong and connected IP system for creators; and
- IPOS also reaches out to a wide array of audiences including the general public, government, and young people, to educate them and raise IP awareness.

IPOS has come a long way since its early days as the Registry of Trade Marks and Patents. It was restructured as the Intellectual Property Office of Singapore in 1999 and became a statutory board in 2001. On 14 February 2014, to strengthen Singapore's patent regime and to make it in line with other established patent offices such as the US, UK, Japan and the European Patent Office, Singapore's Patents regime was amended from a self-assessment system to a positive grant one.

The efforts of IPOS over the years have made Singapore a well-respected location for IP protection, essential for advanced R&D activities. This has helped the EDB and NSTB/A*STAR to attract MNCs to establish their overseas R&D centres such as Procter & Gamble and Continental Automotive in Singapore against keen competitions from other Asian countries. The resultant boost in R&D activities has in turn stimulated IP filing and management services, both local and foreign, to be established and contributing to the growth towards an IP ecosystem.

In addition to supporting local businesses, Singapore's intention is to strategically position itself to service global clients that are exploring growth opportunities in this part of the world. The plan is to professionalise the IP ecosystem in Singapore and around the region with a comprehensive development framework known as the IP Competency Framework (IPCF). Specifically, IPCF maps out competencies needed and projects suitable pathways for a professional career in IP. With the IPCF, interested stakeholders keen on entering the IP industry — fresh graduates, mid-career professionals, training providers — will now have a quality-assured framework to adhere to.

The IPCF sets the standard for five core professional job types in the IP ecosystem, namely the pool of expertise for Patent Agents, IP lawyers, IP Management Directors, IP Strategists, and IP Valuation Analysts. To elevate the overall quality of IP services offered in the ecosystem, IPOS has inked agreements with various professional bodies, including the Association of Singapore Patent Agents (ASPA) and the Institution of Engineers, Singapore (IES), to professionally certify IP Patent Agents/Attorneys and IP Technology

Consultants respectively. This will help deepen IP manpower capabilities under the different IPCF industry clusters.

The R&D community will benefit from the IPCF as there will be more trained IP professionals to help them learn and plan their R&D and innovation strategies. Some R&D scientists and engineers may also seek new career opportunities in the IP ecosystem. As an example, when IPOS formed its first-ever Patent Search and Examination unit in 2013, a number of experienced research scientists and engineers with PhDs joined this unit, making it one which has the largest percentage of qualified patent examiners in the world!

The IP Academy, a subsidiary of IPOS, has been positioned to be the core training provider. The Academy will be the principle adopter of the IPCF in all training programmes in close collaboration with the local universities to serve the market demands in Singapore as well as those in the region. IPOS has also started to stimulate more public interest in the economic aspect of IP. It launched a $100-million Intellectual Property Financing Scheme in 2014 to support local businesses to use their granted patents as collaterals for bank loans. It has also collaborated with the World Intellectual Property Organisation (WIPO) to introduce the annual joint WIPO-IPOS IP Awards. They include the award for the R&D organisation with the highest number of Patent Cooperation Treaty (PCT) patent filings, the award for outstanding IP commercialisation, and the award for outstanding technology creation.

IPOS has also set up two additional subsidiaries: IPOS International and IP ValueLab. IPOS invested $50 million in 2012 to build up a patent search and examination (S&E) unit which has become part of IPOS International. The indigenous S&E Unit has received affirmation on its high-quality work from the IP community and achieved ISO certification for its search and examination. The IP ValueLab helps businesses to unlock the value of their IP through collaboration with global and local business partners and the provision of a suite of solutions in IP management, valuation and monetisation. By enabling businesses through robust valuation of their intangible assets, IPOS helps businesses transform ideas into capitalised assets.

Infocomm Development Authority of Singapore (IDA)

Since independence in 1965, Singapore has created a fair number of government agencies with varying degrees of autonomy to administer different strategic areas. These agencies contribute significantly to Singapore's R&D ecosystem through their focus on S&T to carry out their roles. Over time, agencies have also been merged as their roles have shrunk, grown larger or intertwined with new developments in science and technology. Two such agencies that came together as a result of the convergence of information technology and telephony were the National Computer Board (NCB) and Telecommunication Authority of Singapore (TAS). The merger produced the Infocomm Development Authority of Singapore (IDA) whose aim is to grow Singapore into a dynamic global hub and to leverage on infocomm for Singapore's economic and social development. The role of TAS was essentially to manage Singapore's telecommunications strategy and infrastructure, a role that became redundant with the development of digital communications. Today, IDA is shaping a role for itself in the Smart Nation Initiative and placing emphasis on creating and promoting a culture of using technology innovations to create and build new and innovative products and solutions to address new challenges in Singapore and globally. It plays this role through supporting the growth of innovative technology companies and start-ups in Singapore, working with leading global IT companies, and developing IT and telecomms infrastructure, policies and capabilities. (More in Chapter 8)

The National Computer Board (NCB) was established in 1981 to drive the national computerisation movement and develop information technology in Singapore. NCB was chaired by Philip Yeo, then in MINDEF which was developing in-Singapore defence technology. Given that the first computers and even the Internet were originally developed for defence purposes in the West, the genesis of the national computerisation movement is clear. Its stated mission was to "drive Singapore to excel in the information age by exploiting IT extensively to enhance Singapore's economic competitiveness and quality of life". Its primary goals included promoting the use of IT in private sector organisations, coordinating national schemes to train IT specialists, and implementing the computerisation of the civil service. Information technology can play a role in enhancing manufacturing competitiveness by cutting operational costs and adding value to products and services.

In capability development, the Critical Infocomm Technology Resource Programme (CITREP) was established by the IDA in 1996 with the objective of accelerating the development of emerging, critical and specialised infocomm skills to meet Singapore's infocomm manpower needs. CITREP supports courses and certifications from a broad range of critical and emerging areas, including focus in building up competencies for Smart Nation in the areas of Software Development, Data Analytics, Cloud Computing, Infocomm Security. There are currently more than 100 courses and certifications that support the development of entrants to experts in the ICT industry.

The aim is to deepen the capabilities of infocomm professionals and to ensure that Singapore's infocomm manpower continues to remain relevant to developments in the industry globally and within Singapore. To ensure Singapore's continued leadership in infocomm, IDA identifies strategic info-comm technologies and works with the relevant players in industry and government in areas such as policy, regulation, manpower development, technology pilots, and trend mapping. IDA engages industry, academia, end-users and other stakeholders in the infocomm ecosystem to identify signifi-cant trends in the infocomm landscape and develop an Infocomm Technology Road Map that charts the national infocomm blueprint for the future.

In 2005, a high-level steering committee convened to spearhead the development of Singapore's 10-year master plan to grow the infocomm sector and to use infocomm technologies to enhance the competitiveness of key economic sectors and build a well-connected society. The Intelligent Nation (iN2015) master plan is a living blueprint, jointly developed with the people and the private sector to navigate Singapore's exhilarating transition into, to use IDA's tagline, "An Intelligent Nation, A Global City, Powered By Infocomm". iN2015 aims to fuel creativity and enable innovation among businesses and individuals by providing an infocomm platform that supports enterprise and talent. It serves to connect businesses, individuals and com-munities, giving them the ability to harness resources and capabilities across geographies.

Under the sixth and latest Road Map, IDA expects to see new develop-ments arising from the confluence of innovations that will bring about greater convenience for end-users and the emerging ecosystem of hardware, software and systems-level players that exploit sentient spaces. There will also be tre-mendous opportunities for the infocomm industry, and multidisciplinary research around infocomm will become the key focus of R&D. IDA will con-tinue to develop industry-relevant programmes that bridge R&D labs and industry, to help Singapore keep abreast of future directions and trends, iden-tify business opportunities and strengthen its competitive advantage.

One approach IDA has taken is to create IDA Labs — physical spaces for individuals, companies and government agencies to collaborate. It enables them to work on generating new ideas, developing new technologies and test-ing out proof-of-concepts. IDA Labs aim to focus on:

- Building Singapore-based tech companies and developing local Tech Talent. The aim is to support talented individuals and companies to generate innova-tive intellectual property from ideation to productisation through the lab resources and the mentor and user networks;

- Fostering a culture of creating and building. The aim is to ignite the youthful passion for technology, particularly in experimenting with and building on technology; and
- Strengthening internal capabilities. The aim is to further strengthen the technology capabilities of IDA as well as support innovative technology deployment within government agencies.

Building on the achievements of iN2015 is the private sector-led Infocomm Media 2025 Masterplan (IMM), which seeks to establish in Singapore a Globally Competitive ICM Ecosystem that enables and complements Singapore's Smart Nation Vision, creates Enriching and Compelling Content, and brings about Social and Economic Transformation. The stronger ICM sectors will empower Singapore to tap into the potential of ICM to tackle national challenges, and bring about sustainable and quality growth, as well as promote a better quality of life for Singaporeans.

Public Utilities Board

Another agency that has been added to the story of Singapore's R&D development has been the national water agency, — Public Utilities Board (PUB). PUB manages all aspects of the water cycle in an integrated manner, from sourcing, collection, purification and supply of drinking water, to the treatment of used water and its reclamation into NEWater, as well as the drainage of storm water. This allows PUB to identify and consider the gaps in the water loop, and look towards R&D to provide innovative solutions. "PUB is unique in the sense that we have oversight of the entire water chain from planning to design, from implementation to operation, maintenance and sale of the water. It is all centralised in PUB, which you don't find in other areas like telecommunications or information technology," says PUB Chairman Tan Gee Paw in PUB's commemorative publication on water, *Creating Value: Singapore's Water R&D Journey*.[4] "We do everything from A to Z, and that's why we know what is required, we know the technologies that are available, how to assemble them together and how to get things done." Tan Gee Paw formulated Singapore's first Water Master Plan in 1972.

Water safety and security is a national issue, one that has made the PUB an important agent for change in the R&D landscape. Speaking at the 2014 official opening of Tuaspring Desalination Plant, Asia's largest seawater reverse-osmosis desalination plant and Singapore's second desalination plant, Prime Minister

---

[4]Creating Value: Singapore's Water R&D Journey. Published by PUB, 2013.

Lee Hsien Loong said: "They promoted R&D to develop new sources of water, researching recycling water and desalination way back in the 1970s. Then, the costs were still too high; we did not implement it yet. But we kept an active interest and when membrane technology progressed and become cheaper and more sophisticated in the 1990s, PUB's earlier work enabled it to launch NEWater and to implement desalination on a large scale. And so we developed our water industry, partnering researchers and industry to conduct joint R&D and growing the capabilities of our water companies which now operate in two dozen countries around the world." As Singapore's water demand is expected to double in the next 50 years, ensuring water sustainability becomes ever more vital. R&D thus becomes crucial in ensuring a safe and sustainable water supply for the future as Singapore looks towards more cost-efficient and effective ways of producing water, and enhancing system efficiency.

In July 2006, the Research, Innovation and Enterprise Council identified the water industry as a strategic area where a focus on R&D had the potential to drive economic growth. The National Research Foundation committed $330 million towards public and private sector water-related research in 2006, with this fund later growing to $470 million in 2011. To manage and administer the research fund, the Environment and Water Industry Programme Office (EWI) was set up by the Ministry of the Environment and Water Resources as an inter-agency body. The EWI secretariat aptly sits within PUB's Industry Development Department whose objectives seamlessly overlap with EWI's key strategic thrusts of cluster development, internationalisation and capability development (including both technology and manpower development). With PUB taking the lead, other agencies including the EDB, International Enterprise Singapore (IE Singapore), SPRING Singapore, Workforce Development Agency, JTC Corporation, Ministry of National Development, National Environment Agency, as well as academic partners such as NUS and NTU, and A*STAR, together with industry partners such as Black & Veatch, Meidensha, and GE Water all lend their respective areas of expertise toward EWI's strategic goal of spearheading the development of Singapore's water industry.

EWI has a comprehensive Technology Road Map to guide its R&D efforts which are driven by the strategic objectives of lowering chemical usage, reducing energy consumption, minimising waste generation, improving water quality, decreasing water production costs, and increasing water resources. The key technology domains in this road map include:

- Biological processes;
- Network management;
- Watershed management;
- Desalination and water recycling;

- Sludge and brine management;
- Sensors and instrumentation;
- Underground options;
- Decentralised water treatment;
- Industrial water technologies; and
- Automation, robotics and artificial intelligence.

This road map is continuously updated to meet Singapore's needs as well as to address future global water challenges.

The roles of PUB and EWI complement each other to complete the spectrum of water R&D in Singapore. PUB's operations-centric mandate sees its focus on downstream R&D such as refining a solution through pilots, demonstration studies and test-bedding. While fundamental research is long-term and experimental in nature, the benefits of encouraging and sustaining it can be great given its potential for producing breakthrough technologies and expanding knowledge bases. This is where EWI comes in to support upstream R&D through a number of conduits. These include aiding the development of proprietary ideas at their conceptualisation stage and seeding the innovation of water-related products, processes and applications through funding initiatives such as the Incentive for Research and Innovation Scheme (IRIS).[5] EWI also assists start-ups to precipitate their formation and growth, with the overall goal of seeing an idea through basic research, applied research, test-bedding and the end-point of commercialisation.

As former PUB Chief Executive Chew Men Leong elaborated in *Creating Value: Singapore's Water R&D Journey*: "As we engage in R&D, it has a catalytic effect on encouraging the build-up of more R&D capabilities, not just within PUB but also with all our stakeholders and industry in Singapore. We open our plants for testing; we are involved in R&D in search of solutions not just for ourselves but also for the whole water sector. Suddenly, you see R&D growing beyond its strategic value and taking on an economic value as we build up the water industry." He added: "While we were in the process of reaching out, we realised we could build a network. We could encourage companies to come to Singapore and undertake R&D, to set up R&D centres, and slowly, build up an R&D ecosystem. Over the last decade, we have built a sizeable R&D ecosystem that has economic values, where R&D projects can create products for various other markets. At the same time, some of these activities can be ploughed right back to create solutions for PUB."

---

[5] Environment and Water Research Programme–IRIS is an existing initiative used to support funding of both basic and applied R&D projects which possess recognisable potential in developing innovative and novel solutions in key technology domains identified in EWI's Technology Road Map.

AKA Singapore Incorporated

The huge investments in R&D that begun in 1991 are about building capacity, capability and being future-ready. It has developed to what it is today by leveraging on the core principles that helped to bring about economic progress and develop the nation. Identifying the outcomes and putting in place the strategies to realise these outcomes have not been just a matter of luck, says Dr Raj Thampuran, Managing Director of A*STAR. The element of luck comes in the geographic position of Singapore but it was the right policies executed by people that produced the Singapore success story and the story of Singapore Inc. Says Dr Thampuran: "We can bring many things together. Whether it's across public sector, whether it's across performers, across ideologies, whatever it is, our ability to bring things together gives us a compactness and advantage that we should not take for granted. That's been key. If Singapore is committed to something, we are able to bring many things together that other countries find more difficult to do either because you have to traverse geographies, you have to traverse doctrines. In Singapore, we are able to move with some speed and the fact that there is some coherence and integration means that what we are doing we can do it all in one place and bring in all the components that are needed. I think we have always been committed to this idea of capabilities. 'Capabilities' is almost a furtive concept but it is basically the abilities of people. Developing capabilities is an investment in making people more skilled, understand things more deeply, make smarter choices. We have also never shied away from capacity building. If we have to invest in certain infrastructures, we did it. So long as we knew how it contributed towards an integrated plan. That was important. We knew that laws would be important, so IP laws were put into place with stringency in enforcement. It is a distinct advantage, a comparative advantage. Our respect for IP, our ability to prosecute is considered of great value to companies because they know they will be treated fairly and there will be justice here. That is important. It is not just all these things. This is a story of Singapore Inc, about Singapore being a place where talent would want to come. So then it became this whole idea of being a lot more contemporary in the cultural scene. Being accepting of people from different backgrounds and cultures, trying to have things in our society that give a quality of life for those whom we attract to live and work here wherever they come from. Our all-embracing, if you like, way of accepting the best talent from wherever they come from and porosity. So I cannot say these were the few things. It's actually a collection of many elements that you bring together which then becomes the Singapore Inc story."

Chapter 4
Research in Physical Sciences and Engineering

Hang Chang Chieh and Raj Thampuran

In 2001, there was a change of leadership in NSTB, with Philip Yeo, chairman of EDB, taking over from Teo Ming Kian who then became chairman of EDB. Yeo called for a reorganisation of NSTB for a more targeted focus on R&D: he wanted to develop Biomedical Research but he also wanted to attract more industry players to the Science and Engineering R&D landscape. Until 2001, NSTB was basically a planning and funding agency. All the mission-oriented research institutes (RIs) and centres that had been set up with NSTB funding were being managed by the universities. Consequently, there were overlapping research objectives in some of the research centres. Yeo decided that it was time for NSTB to take over active management of the RIs as well as further strengthen their links with industry. Thus, Prof CC Hang was persuaded to become full-time Executive Deputy Chairman of the renamed NSTB, now A*STAR, with the task of rationalising the engineering RIs, grouping them, through the coordination by a Science and Engineering Research Council (SERC). The concept of R&D management through councils was similar to the structure of the UK's Engineering and Physical Sciences Council. Like each of the RIs, the Council would have a panel of international advisors from universities, research institutes and industry leaders. In parallel, Yeo himself was driving the start of the Biomedical Sciences (BMS) Initiative and the formation of the Biomedical Research Council (BMRC) to coordinate the setting up of new research institutes to supplement the role of the Institute of Molecular and Cell Biology (IMCB).

The engineering research landscape at the time consisted of institutes whose origins could be traced to 1981, although not as research institutes. Thus, the national computerisation drive that began in 1981 had much to do with the formation of the Institute of Systems Science (ISS) also in the same year. In 1981, the Singapore Government had announced the formation of the National Computer Board (NCB) headed by Yeo who was then with MINDEF and involved with the then nascent DSO National Laboratories. NCB was to

Singapore Institute of Manufacturing Technology: Some High Points

- SIMTech applied the developed ultrasonic vibration-assisted cutting method to directly cut steel-based alloys to produce optical moulds for use in various industries. The technology is licensed to Delta Optics which remains the only commercial company worldwide with the capabilities to machine hardened steel directly to produce mirror surface finishes better than 8nm Ra. Since its incorporation, the company has developed an international clientele.
- Together with Philips Consumer Lifestyle, SIMTech co-developed wet chemical coatings based on sol-gel chemistry. The SIMTech-Philips team was awarded the National Technology Award in 2002 for the novel combination of aluminium anodising and sol-gel coating for iron soleplates, a breakthrough that enables Philips to maintain leadership in high-end irons.
- SIMTech developed a novel combination of 3D inkjet printing process with conventional processing methods using FDA-approved biocompatible polymers to fabricate bone tissue scaffolds.
- SIMTech developed the Liquid Forging technology to fabricate intricate and structurally-superior components that are not attainable with traditional die casting and extrusion. The technology is ideal for components that require thin wall widths and high aspect ratios to minimise their weight and materials. A winner of the National Technology Award in 2008, Liquid Forging has been successfully licensed to companies to produce

(Continued)

lead the national computerisation drive starting with the civil service and together with EDB, to develop the computer services industry. Legislation was passed and NCB was set up in September 1981. In the 1980s, IBM was a leading mainframe manufacturer in the world although it did not have manufacturing facilities in Singapore. It was going to be the biggest supplier of mainframe computers to the Government and so was persuaded to help NUS start up the Institute of Systems Science (ISS). NUS started its computer science and information technology degree course in 1983. In 1986, NCB created the Information Technology Institute (ITI) as its applied R&D arm for developing advanced IT innovations in Singapore, with Lim Swee Say as its founding director. ITI was subsequently merged with the research group in ISS in 1998 to become Kent Ridge Digital Labs (KRDL), with a balanced focus on research in both software and hardware in information technology. KRDL would be responsible for spinning out a substantial number of commercial

(Continued)

light-weight and high-performance products for thermal management and structural applications.

- SIMTech collaborated with four SMEs from the precision engineering industry to build up their capabilities in mould design and fabrication, metal insert and plate fabrication, injection moulding process control, spray coating and assembly processes through a Collaborative Industry Project. The SMEs were audited and approved as SIA Engineering Company's service providers, enabling them to migrate seamlessly into the aviation industry business.

- SIMTech collaborated with a local SME, Film Screen, to implement large-size printed lighting on vehicles, opening up new lighted advertisement options in Singapore. The Large Area Processing technology enables Film Screen to achieve the world's first 12-sq-m continuous printed lighting film on a double-decker bus as a medium of advertisement for brand enhancement. In another usage, the technology is also applied to warm up refrigerated blood before use.

- A microfluidic-based droplet generator has been developed by SIMTech whereby gas bubbles agitate the fluid stream in a microfluidic chip to form droplets. The microfluidic chip is disposable and can be sterilised to eliminate contamination. The throughput is high, >1000Hz, compared to emulsion droplet formation. Microbeads have been produced from biocompatible polymers and cells encapsulated can be cultured as usual. This technology can be applied in tissue engineering and in the pharmaceutical industry.

start-ups that formed part of the early history of technology commercialisation in Singapore. These start-ups in the late 1990s were part of the dot.com era and provided a wealth of experience for the newly-developing patent legislation and patent profession.

Another engineering institute that would become part of the Singapore Institute of Manufacturing Technology (SIMTech) was started essentially as a training institute in 1985, and this was the Grumman International-Nanyang Technological Institute for CAD/CAM (GINTIC). Grumman International, an aircraft manufacturer, had sold some sophisticated surveillance planes to the Singapore Armed Forces and the establishment of the institute with assistance from Grumman's engineers had been a quid pro quo negotiated by Singapore. GINTIC would subsequently change its name to GINTIC Institute of Computer Integrated Manufacturing. Meanwhile, an Institute of Manufacturing Technology had been one of the two advanced engineering research institutes proposed in the

Institute of Microelectronics: Some High Points

- Bluetooth standard was released in 1999 to address the need to replace cable connections with wireless for data transfer between portable and fixed devices in close proximity. In 2000, IME and OKI jointly developed one of the earliest Bluetooth chips operating at 2.4 GHz realised by mainstream compatible manufacturing processes. The highly integrated Bluetooth RF transceiver was designed on 0.35 μm CMOS RF technology, providing a compact and low-cost solution. The breakthrough outcome was presented at the International Solid-State Circuits Conference 2001, the first time Singapore was represented at this prestigious conference.
- In 2006, IME developed a platform technology for fabricating highly manufacturable and cost-effective CMOS silicon wires of nanometer dimension as thin as 3nm, which expanded the possibilities of device miniaturisation for biomedical, data storage, energy harvesters and opto-electronics applications. The technology was applied successfully to a novel transistor design that addressed the limitation of Moore's Law while enabling functional high-speed logic circuits and memory cells at low voltage. This groundbreaking work won IME both international and national accolades, including the 2007 IEEE George E. Smith Best Paper Award and the 2008 National Technology Award.
- In 2007, IME developed a UHF Radio Frequency Identification (RFID) reader chip for EPC Gen2 standard. Compared to the conventional RFID reader that measured 6 × 6 inches in size, weighing 1–2 kg costing a few thousand dollars, the new RFID chip cost less than $100 and weighed less than 100 grams with a dimension about the size of a name card. This breakthrough enabled the elimination of bulky and costly reader modules, and paved the way for wider market adoption. The success of this reader chip led to a spin-off company, FeRmi.
- In 2010, IME, in collaboration with the National Heart Centre Singapore, developed a microfluidic system that can be used as a bedside tool for point-of-care diagnostics to assist clinicians in detecting the presence of rare

(Continued)

1990 DPC Paper. Thus, in 1993, this proposed Institute of Manufacturing Technology was merged with GINTIC to become the GINTIC Institute of Manufacturing Technology (GIMT), with Dr Frans Carpay as executive director. In 2002, during the reorganisation of the engineering RIs, GIMT became SIMTech.

(Continued)

cardiac biomarkers which can promote blood vessels repair. The microfludic system enabled blood samples to be placed into the microfluidic chamber with minimal cell loss, followed by extraction and conversion of the biomarkers into electrical responses for analysis. The highly sensitive device is capable of detecting a high concentration of EPC (Endothelial Progenitor Cells) in 100 μl of whole blood. In addition, it is able to conduct sampling and analysis in less than one hour compared to conventional requirements of four to five hours. This is critical for enabling efficient prescription of treatment to heart patients.

- IME and Cubic Micro announced in 2014 that they had developed and demonstrated a 400 MHz radio frequency (RF) transceiver with the highest power efficiency and leading performance reported to deliver high quality signals over industry's widest coverage in wireless sensor network applications. The transceiver is integrated with a highly configurable baseband that allows users to customise transceiver performance for specific applications ranging from wireless smart energy management and security control in homes and buildings to long-range remote industrial monitoring.

- As early as 2006, IME developed key process technology and device building blocks that are compatible with industry practices and standards, with the added advantage of being CMOS-compatible for foundry manufacturing. The capabilities developed have successfully translated into the earliest prototyping of fully functional and high performance optical devices such as transceivers and photodetectors. These technological breakthroughs won IME the prestigious President's Technology Award 2010 and the IEEE George E. Smith Best Paper Award 2007. In 2013, leveraging on IME's established silicon photonics capabilities, it came up with the world's first 40 Gbps silicon-based optical modulators based on multilevel modulation with Fujikura. The 40 Gbps modulator is based on a differential quadrature phase shifting keying that is capable of doubling the rate of information transmitted over a single communication system compared to the on-off keying format widely used at the time. IME eventually further pushed the speed up to 100 Gbps.

The other advanced engineering research institute proposed in the 1990 DPC Paper had been the Institute of Microelectronics (IME) which was founded in 1991 with Dr Bill Chen as its executive director. IME's mission was to add value to Singapore's microelectronics industry which had been identified as a growth industry after 1986. The institute aimed to develop competencies in industry-relevant areas, innovative technologies and a body

Data Storage Institute

The EDB was instrumental in starting up the Magnetics Technology Centre that was later renamed Data Storage Institute, today a leading research institute in data storage. Recalled Prof Low Teck Seng in a 2011 interview: "In 1990, the EDB folks came to see me and I was a young senior lecturer at NUS and they said 'We want to establish for ourselves in Singapore a research institute very similar to the ones that exist in the United States'. The top institute in the United States at that time was the Data Storage Systems Centre at Carnegie Mellon University. There was also a very good centre in San Diego, in UC-San Diego. There was also a very good institute in Minnesota called MINT (Centre for Micromagnetics and Information Technologies), and then there was also a very well-known institute in Santa Clara University. So, they gave me these four annual reports and they said 'We need to establish something like that'. And the reason why they came to me was because I was probably the only person in NUS at that time doing something related to magnetics." The EDB proposal had come at the tail end of the Science Council and NSTB had not yet been established. There was no funding mechanism for something like that, nor was it easy to make the necessity for such an institute understood by the traditional funding sources. Pushed by EDB chairman Philip Yeo, who was insistent that such an institute be founded, Prof Low went ahead to make presentations and solicit funding. With the establishment of NSTB in 1991 and the announcement of the first National Technology Plan, funding for the Magnetics Technology Centre was no longer an issue.

Founding executive director Prof Low grew the original 20-man R&D team into a 180-strong institute with world-class research capabilities. He recalled his initial difficulty in starting the institute: "Building the institute from scratch was tough. It was difficult, but along the way there was great excitement." Setting up and developing such an advanced engineering institute showed that there was local talent that was up to such a task.

of intellectual property while at the same time growing a talent pool that could be transferred to industry partners and to move SMEs up the technology ladder. All these institutes had been started with international research leaders and talent.

One advanced engineering research institute that was started up with local talent was the Data Storage Institute (DSI), today one of the leading research institutes in data storage in the world. One reason for turning to local talent was that there were few foreign resources in this field to tap into, data storage being still relatively new. Data storage technology was so new that the bulk of advanced research was being done in the disk drive companies with the pioneers, Seagate and IBM, doing far more advanced research than the US universities. The Magnetics Technology Centre was started in 1992 by Prof Low Teck Seng of NUS with the aim of developing core competencies in technologies related to data storage to add value to the disk drive manufacturers already in Singapore. At its peak in the 1990s, Singapore was producing over half of the world's hard disk drive units. Another longer-term goal was to develop the technological capability to assist not only existing industry but also to encourage and cultivate spin-offs. In 1996, the institute filed its first patent for its fluid film bearing motor prototype. By then the Centre had expanded its scope from an initial focus on mechatronics and electronics to encompass magnetic read sensors, magnetic media and optical programmes. So in 1996 the Magnetics Technology Centre was renamed Data Storage Institute to better reflect its enlarged research portfolio, said Prof Chong Tow Chong who became Executive Director of DSI after Prof Low. Prof Chong's research interests at that time had been in optics and laser technology.

The Institute of Chemical and Engineering Sciences (ICES) had its origins in NUS. In 1994, a group of academics from across disciplines started a research group to work with industry. It was named Process Analysis and Optimisation Enterprise (PAOE) and had a part-time director, Prof Ching Chi Bun of the Chemical Engineering Department. The group had started up at the suggestion of Prof CC Hang who was then the (part-time) Deputy Chairman of NSTB and the first Director of Research at NUS. Prof Hang had worked in Shell Petroleum prior to becoming an academic at NUS. Thus, he understood the challenges of large petroleum companies and observed that the entrepreneurial start-ups in the chemicals industry were mostly in the area of specialty chemicals. In 1998, with both Ministry of Education and NSTB funding and EDB's advisory support, PAOE was upgraded to a research centre within NUS. Renamed the Chemical and Process Engineering Centre (CPEC), its first full-time Director was Prof Ching Chi Bun. Because of the growing chemicals sector, Prof Hang was able to get NSTB to support the research work of CPEC and eventually upgrade it to the status of an NSTB Research Institute. When such a decision was made around 2000, an international search for a Director was conducted and Dr Keith Carpenter from the UK was appointed the founding director of the Institute of Chemical and Engineering Sciences (ICES). In 2004, ICES was relocated from NUS where it had facilities in a basement of the Engineering Faculty to purpose-built facilities on Jurong Island to

Institute of Chemical and Engineering Sciences: Some High Points

- In the field of sustainable chemicals, ICES has made several inventions which increase the energy efficiency and reduce the carbon footprint of chemical and pharmaceutical processes, and reduce dependency on traditional petroleum-based feedstock. Some of the inventions include a stable and selective solid acid dehydration catalyst to produce olefins such as ethylene from bio-alcohols such as ethanol, butanol and mixtures and a novel catalyst to produce aromatics from bio-ethanol.

  ICES also improved biogas production from municipal waste and developed sustainable processes to produce adipic acid and acrylic acid from lignocelluosic agricultural waste such as wood, empty palm fruit bunches, sugarcane bagasse or rice straw. ICES has also developed a novel process to capture CO_2 to produce materials for landfill and construction.

- In the field of pharmaceuticals, ICES has contributed novel formulations which increase our toolkit for treatments and reduce the cost of drug production. ICES has also developed a novel, highly active pharmaceutical formulation based on powders as well as a new drug delivery system. These have potential applications in extending patient life and improving drug performance in the pharmaceutical industry.

- In the field of specialty chemicals, ICES invented a sugar-based surfactant micro-emulsion containing essential oils for cosmetic and pharmaceutical use. It is a good agent for topical treatment that has excellent thermodynamic stability and biodegradability. At the same time, ICES also invented a fire-retardant polymer and biocide compounds of anti-fouling solutions. The work done by ICES has attracted companies such as Syngenta, Mitsui and Dystar to have lab-in-RI arrangements with ICES which could potentially grow to become fully fledged R&D Centres in Singapore, providing good jobs for Singaporeans.

be a research partner to the chemical companies cluster there. Minister for Trade and Industry Lim Hng Kiang said at the official opening of the institute that year: "The idea of establishing a research institute in Jurong Island was mooted when the island was officially opened in 2000. It is part of a conscious effort to create and sustain value for the chemical and petrochemical companies who have made Jurong Island their home. ICES's proximity to petrochemical companies on Jurong Island and the biomedical companies in the neighbouring Tuas Pharma Park will provide mutual convenience and greater working synergy for both the companies and ICES. Researchers at ICES can work hand in hand with chemical and petrochemical companies in Jurong Island on joint research collaborations to develop new products and processes. Companies in Jurong Island can tap on the state-of-the-art research and training facilities at ICES to meet the demands of the highly knowledge-intensive chemicals industry. Being situated on Jurong Island will also provide greater opportunities for researchers to meet informally, exchange ideas and build a close rapport with their industry counterparts." In 2009, ICES opened the first pilot-scale research facility in Southeast Asia for solving the problems of scale-up and manufacturing in the pharmaceutical and petrochemical industries and to develop new process research techniques.

Another research institute that started with local talent was the Centre for Computational Mechanics (CCM) in NUS. It was started in 1994 by Prof Lam Khin Yong whose PhD thesis at MIT in the 1980s had been on developing a 3D computational model for hydraulic fracturing, commonly known today as fracking. As an academic at NUS, his earliest research involvement had been with the Republic of Singapore Navy arising from his research focus on computational mechanics. At the time, the Centre for Computational Mechanics (CCM) had received one of the biggest research grants of some $7 million, said Prof Lam. In 1998, CCM was merged with the National Supercomputing Research Centre (NSRC) that had originally been formed in 1993 to promote and support the use of high-performance computer applications. The institute thus formed was named the Institute of High Performance Computing (IHPC) with Prof Lam as its founding executive director. The institute research spanned solid and fluid computational mechanics, computational chemistry, virtual product design, digital modelling and visualisation and advanced computing technologies.

One of the more complicated mergers was the creation of the Institute for Infocomm Research (I^2R). It brought together centres that had grown out in NUS and NTU, and merged them with NSTB's Kent Ridge Digital Laboratories (KRDL), which itself had been formed earlier by merging the research groups in NUS's Institute of Systems Science and NCB's Information Technology Institute. From NUS, there was Centre for Wireless Communications and from NTU, there was Centre for Signal Processing. Prof Hang, who oversaw these series of mergers, recalled the difficult process of convincing the various research entities to work together to achieve synergy: "Finally, I had to give it a nice and powerful name. Institute for Infocomm Research was chosen as I^2R is the formula of Electrical Power." Today, I^2R has

Institute of High Performance Computing: Some High Points

- IHPC, together with the National Heart Centre, Singapore, has devised a cardiac diagnosis method that generates comprehensive spatio-temporal indices to describe regional heart ventricle morphology and function. This new approach allows heart patients to be correctly diagnosed where conventional diagnosis methods fail.
- High speed optical telecommunications are affected by the spreading of light pulses, or chromatic dispersion. This in turn causes errors and loss of information. In order to solve this, chromatic dispersion compensators are used where low power consumption is typically required. IHPC has developed the first 3D electrical and optical simulation of chromatic dispersion compensators. This allowed our industry partner to reduce device power consumption from microwatt to nanowatt regime.
- Development of fully stretchable and reliable electronic circuits with researchers at the University of Illinois, Urbana-Champaign, and Northwestern University in Chicago. Stretchable electronic circuits are achieved by depositing stretchable electronic devices and circuits onto stretchable substrates or by embedding them completely in a stretchable material such as silicones or polyurethanes. These devices enable a paradigm shift for the architecture and design of flexible and curvilinear electronics with exceptional mechanical characteristics and functionalities. Simulations by IHPC have provided new insights into the mechanical characteristics of stretchable silicones with micron-size wavy geometries and have added a new dimension to the properties of silicones and opened up new applications such as electronic eyes, bendable LCS screens, surgical gloves

(Continued)

over 580 researchers working in a spectrum of areas to advance Infocomms R&D.

After the formation of these engineering research institutes, Prof Hang then helped to put up a research institute in materials to support the needs of both the research institutes and the industry. The Institute of Materials Research and Engineering (IMRE) attracted a Singaporean, Prof Shih Choon Fong, who had developed an outstanding career in the US, back to Singapore to serve as its founding executive director. Established in 1997, IMRE's research includes organic solar cells, nanocomposites, flexible organic light-emitting diodes, solid-state lighting, microfluidics and next-generation atomic scale interconnect technology.

(Continued)

with intelligent sensors, prosthetic limbs that can alter their shapes in response to temperature or pressure changes, and wireless medical devices.

- High energy consumption in water treatment is a big concern for industry. The complex fluid flow and gas aeration involved in the treatment process is the key factor for energy usage. A better hydrodynamic performance will result in energy saving by high energy efficiency. Improvements by IHPC in water filtration membrane technology using computational fluid dynamics tools and simulations have provided insights into a complex and large-scale multiphase flow phenomena with membrane filtration effects and arrived at an optimal system that increases productivity and efficiency. The new optimum designs based on the simulation results have been adopted by a few water treatment plants in the US and Middle East.

- Viewing glasses are used in many applications. However, they are limited in their viewing angles and polarisation of light. As such, there is a demand for viewing glasses that can meet these two critical requirements. IHPC's design of photonic crystals with light reflection for all angles and all polarisations have wide applications such as ultraviolet filters for sunglasses, safety glasses, visible light blocking for security and defence applications, and infrared and UV photography filters.

- IHPC has developed and implemented an urban microclimate modelling tool to optimise the urban design to mitigate urban heat island effect and climate change impact. The tool can be used to simulate microclimatic conditions to achieve better performance with various urban factors, including designing buildings with improved layout to enhance natural ventilation; optimum land usage for roads, greenery, buildings, and water bodies; new materials and ventilation systems to reduce energy usage.

Following the reviews and mergers, the number of research institutes to be coordinated by the Science and Engineering Research Council (SERC) was reduced to seven plus one centre. The hardest part was getting people to give up territory and to see the advantages of larger institutes, thereby not dissipating scarce RSE talent in duplicating work. SERC was now aligned with Singapore's four main industry sectors: electronics, infocomm (information and communications technology), chemicals, and engineering. SERC has strengths in key areas: manufacturing technology, computational and device technologies, infocomm technology, communications and media, chemistry, mechatronics and automation, and metrology. The research institutes help these industry sectors to meet the challenge of innovating and

Institute for Infocomm Research: Some High Points

- I²R's Snap2Tell, a set of robust image recognition SDK (Software Development Kit) that identifies pictures captured by mobile phone cameras, and further provides augmented information, has won many awards, including the 2014 IES Prestigious Engineering Achievement Award. The technology has been licensed to a dozen SMEs and is deployed by I²R's start-up, Knorex, in major newspapers in seven countries. It was also deployed as a productivity tool to benefit SMEs. For example, Manja Delights', a hawker stall, uses Snap2Tell to take online orders; Feinmetall, an engineering firm, uses Snap2Tell as a training tool to improve its training productivity by 30%.
- I²R's BreathOptics™ Technology, based on Fiber Optic sensing technology, detects normal breathing cycles, movements and special events such as abnormal breathing patterns in new-born babies. The technology was licensed to ExcelPoint which developed an intelligent baby monitoring mat product that was test-bedded and validated at Singapore's KK Women's and Children's Hospital. The innovation (http://www.ibabyguard.com/) won several awards including Best Innovation For Parenting (Thailand's *Real Parenting*), Hong Kong Toys & Baby Products Award 2012 and 2013, and was a Finalist in the Asian Innovation Awards 2011.
- I²R's Abacus language engine achieved a consistent leading performance in US National Institute of Standards and Technology (NIST) international benchmarking competitions. The Abacus engine is able to address unique problems that Asian languages face, such as multilingual speech and tonal language processing, and translation between Asian languages. The Abacus engine has been licensed to more than 40 companies. In particular, it was adopted by Baidu in 2012 to power the Lenovo A586, the world's first voice-print smartphone. For these contributions, the team received the President's Technology Award in 2013.
- I²R's data analytics technologies won first place in the 2013 General Electric Flight Quest Challenge competition, out of 180 international teams. The technology is aimed at helping GE make flights more efficient and punctual. I²R's data analytics algorithms beat flight time predictions from expert air traffic controllers by 40%.

(Continued)

upgrading technology to stay relevant and be at the forefront of technologies. As more countries in the region began industrialising, the easier options of simple manufacturing or assembly work moved away to countries with cheaper costs while the technologically more advanced work has remained.

(Continued)

- Advanced Audio Zip (AAZ) is the first "made-in-Singapore" technology adopted by normative MPEG international standards. AAZ is capable of packing any music file into a compressed file less than half its original size with no loss in quality during playback known as lossless quality. AAZ was adopted as the worldwide reference model in MPEG scalable to lossless audio workgroup and published by ISO as an international standard in June 2006. The team received the National Technology Award in 2007.

- I²R's armoured-cable-based Fibre Bragg Grating (FBG) Perimeter Intrusion Detection System was first deployed around Singapore Changi Airport's 22-km perimeter. The patent-pending FBG technology and unique intelligent adaptive signal processing algorithm guarantees high sensitivity, extremely low false alarm and nuisance alarm rate. It was benchmarked and outperformed market leaders in 2010 in all intrusion event categories. It was licensed to ST Electronics (Satcom & Sensor Systems), and has since been deployed to sites locally and overseas.

  Together with ST SatcomS and Changi Airport Group, the technology won both the ASEAN Federation of Engineering Organisation (AFEO) ASEAN Outstanding Engineering Achievement Award and the IES Prestigious Engineering Achievement in 2013.

- I²R embarked on a research programme REVIVE (Reverse Engineering Visual Intelligence for cognitiVe Enhancement, http://www.i2r.a-star.edu.sg/revive/index.html) with visiting investigator Prof Tomaso Poggio from MIT and CBMM (http://cbmm.mit.edu/) to develop Augmented Visual Intelligence framework and tools, from understanding human vision and memory to practical assistive technologies on wearable devices. Among the outcomes after one year, the team has developed a prototype intelligent glasses that can aid the elderly in their recognition of faces and loved ones, and also remind the users to take their medication at the appropriate time. The application is based on some of the visual intelligence technologies developed under REVIVE. The prototype was featured on the cover of the Home section of *The Straits Times* on 10 July 2015. It is envisioned the technology will enable more elderly people to cope and ease the load of caregivers in ageing Singapore in future.

The presence of the research institutes and their relevance to industries have kept MNCs in Singapore, and SMEs engaged here. Data storage and semiconductors continued to make important contributions to the Singapore economy up to the end of the first decade of the 21st century. The SERC

Institute of Materials Research and Engineering: Some High Points

- In 2002, IMRE developed composition-tunable quantum dots which are stable, monodispersed semiconductor nanocrystals alloys that can be used as biological labels for biomedical research.
- In 2002, IMRE developed a novel nanoimprinting and fabrication technique that can be used to form nanostructures on flexible polymer substrates. The innovative approach allows the formation of multi-layer and 3D nanostructures and creates new platforms for industry processes and applications. IMRE subsequently formed a consortium in 2010 to further develop this technology for industry applications, and the Nanoimprint Foundry was formed in 2013.
- In 2003, IMRE developed a composite material that can be used as nano-magnetic tags for identification of valuable items to prevent forgery and counterfeiting. This led to the creation of IMRE's first spin-off company, Singular ID, in 2005.
- IMRE developed an Ultra-high Flexible Barrier Film that protects sensitive devices such as organic light-emitting diodes and solar cells from moisture 1,000X more effectively than any other technology available. The inventors have spun off Terra-Barrier Films Pte Ltd in 2009 to develop the technology further.
- In 2014, Tera-Barrier Films Pte Ltd announced the invention of a new plastic film using a revolutionary nano-inspired process that makes the material thinner but as effective as aluminium foil in keeping air and moisture at bay. The stretchable plastic has one of the lowest moisture vapour transmission

(Continued)

research institutes continue to help the electronics and engineering industries meet the challenge of continuous innovation to stay relevant and be at the vanguard.

The Rationale for the Advanced Engineering Research Institutes

Prof Hang had an early introduction to the concept of science parks and the role they play in R&D and building economic relevance in 1981, a decade before he became the founding Deputy Chairman of NSTB. Thanks to his command of Chinese — his primary and secondary schooling was in Chinese — he was asked to accompany Dr Tony Tan, the new Vice-Chancellor of NUS on a visit to Taiwan. He was to act as Dr Tan's interpreter and private secretary. One of

(Continued)

rates (*mvtr*). Owing to its uniquely encapsulated nanoparticle layer, this air and moisture barrier is about 10X better than the transparent oxide barriers that are in use to package food and medicines.

- IMRE'S Translucent Organic Solar Cells which are flexible materials have a wide range of applications from energy-generating tinted windows to powering portable electronics. The material allows light to pass through yet absorbs enough light to generate electricity.

- IMRE perfected a new range of patented microneedles that can be mass-produced more readily and at a lower cost than current microneedles technologies. Micropoint Technologies Pte Ltd, an investment of Japanese conglomerate Sumitomo Corporation Asia, was formed in 2008 to exploit the technology.

- In 2010, IMRE developed flexible composite materials capable of dissipating high impact energy that have potential applications in body armour and protective clothing. IMRE's spin-off company, Sofshell, was formed in 2012 to commercialise this technology.

- Together with Data Storage Institute and Singaporean precision equipment manufacturer Solves Innovative Technology, IMRE built a machine capable of producing nano-meter-sized components in wafer-scale volumes for applications in consumer electronics such as hard disk and optical storage media. This first made-in-Singapore mass-production nanoimprinter is a significant improvement on conventional nanoimprinting processes. Its double-sided imprinting reduces processing time and allows a customised dosing system. It also allows imprinting in a vacuum which prevents air bubbles, thus achieving more accurate patterning.

the places the visitors were taken to was the newly established Hsinchu Science Park that had been set up by Taiwan's former Finance Minister Li Kwoh-Ting in 1980. (Today the world's top two semiconductor manufacturers — Taiwan Semiconductor Manufacturing Company and United Microelectronics Company — are in Taiwan. Hsinchu Science Park also hosts the largest number of 12-inch wafer-producing fabs in the world.[1]) At the time the Singapore visitors did not understand why there was a need for such R&D activities. Singapore attracted MNCs who brought with them their IP and manufacturing technology. Singapore SMEs did the downstream manufacturing and picked

---

[1] Information from Wikipedia retrieved in January 2015 under Hsinchu Science Park.

up the technology that was needed. Taiwan, on the other hand, did not attract MNCs because its political situation was unstable, and so it did reverse engineering coupled with its own R&D, helped by the Taiwanese scientific diaspora who had gone abroad and developed their expertise in the US in particular. Ten years later in 1991, when Prof Hang, as Deputy Chairman of the new NSTB, led a team to look at Taiwan's R&D he recalled: "I went back and got a rude shock. Those small initial investments turned out to be such big technology companies. For example, Taiwan Semiconductor Manufacturing Company was huge. Within 10 years they had done so many things. They were manufacturing PCs, semiconductors, and had built wafer fabs on their own. A lot of their companies such as Acer started and grew rapidly during that time. Their science park turned out to be very applied and innovation-oriented. Their National Research Institutes were spinning off a lot of their activities into the science park. So within 10 years they had built industries of their own. Then suddenly I realised they had to do it because they did not have foreign MNCs. We have foreign MNCs that did all this R&D at home, brought it to us. Therefore we lived symbiotically with MNCs and consequently we didn't know how to grow it. We only knew how to partner them in the downstream activities. We had no experience in doing indigenous R&D, creating new products, and thinking big about new industry creation. All these things were done back at the home of the foreign MNCs. Whereas Taiwan had to develop its own industries with its own R&D because MNCs did not want to go to Taiwan."

Already by the 1990s, simply picking up technology from the MNCs was no longer enough. To stay ahead of the economic curve and to keep MNCs in Singapore, there had to be new areas of research to add value to manufacturing. Said Prof Low, then MD of A*STAR in 2011: "I wouldn't say that the research topics are industry-driven. I would say that, in strategising what A*STAR does, being mission-oriented, we have a fairly clear sight of the economic landscape that we are in, and within that landscape, we will need to have a portfolio of research that needs to be done so that we continue to be relevant, and that requires fundamental research, applied research, and translation of industry research. So, what are the areas that we go into, for example? This is pretty global. We know that in engineering, microelectronics, semiconductors, will be a key driver for many, many more years, but what do we do in this domain that allows us to be relevant to the companies that choose to relocate to Singapore and companies that we want to attract to Singapore, and what technology and science is relevant for the small and medium-sized enterprises that we are going to grow in Singapore?

"The first step is to do a technology scan. We have to map, and then we have to strategise, and the outcome is the direction that the Institute of Microelectronics takes. It also determines the strategies in establishing some of the newer programmes. For instance, there is a big new industry that requires a tremendous amount of electronics, sensors and such. This is the Marine and Offshore industry. So, what are companies in the Marine and Offshore industry interested in? For example, high temperatures because they drill very deep. So, I need to look into programmes which could help them at creating semiconductor systems that could withstand beyond 200°C and harsh environments at depths beyond 3,000 m. Indeed, we have now established a research programme in this area, one that is relevant to this industry and which will attract new companies to locate in Singapore. Another example. As we move forward into the new era where smart power grids become important, the smart management of power becomes critical. Because of that, we have established a new research programme on this. It is very basic work before we can do the applied work, and then finally, working with companies to do translation over into industry. How we strategise is to be constantly aware of the technology landscape, the links to the economic landscape and to craft our economic strategies accordingly. Singapore cannot be engaged in every economic industry area because we are small. So, this is, in a broad sense, industry alignment.

"But if you ask any scientist, we like to see impact from our work. Impact from our work is realised in two different forms. One is publications, in the sharing of knowledge, and this is altruistic, because we push the frontiers of knowledge and science. The other impact, of course, is translation to industry, where it sees real impact on the economy. But there's an altruistic link to it, because finally our work contributes to the well-being of society. But we don't live on love and fresh air. So, finally, if our economy grows, there are resources that flow back to research. For us, it's a closed-loop system we're looking at, and for A*STAR, it sharpens the sense that we are mission-oriented, because we have universities that can focus and concentrate on the basic research. But we do not forsake basic research. In fact, we do basic research, and, in the basic research we do, we actually are able to multiply and leverage, through a leveraging with the universities. So, by working with the universities in areas that we are interested in, we actually expand the amount of basic research that is done, that flows down to us. This is not peculiar to us, except that in Singapore, by being small, things become more obvious. In the things we do, a lot of policies come to a sharper focus because we do it in a much more deliberate fashion. Because of our size, it becomes more deliberate in terms of its articulation."

Fusionopolis

The original name "Technopolis" was renamed Fusionopolis in 2003 to reflect its aim of fostering the fusion of ideas by co-locating the info-communications, media, science, and engineering industries. At the time, the science and engineering research institutes were scattered across Science Park, NUS, and NTU. Prof Chong Tow Chong, then the Executive Director of SERC, was Chair of the Fusionopolis Working Committee. In an oral history interview in 2015, Prof Chong said: "Basically, we wanted to have an integrated capability, and that was to be our unique, differentiating value proposition. Fusionopolis is not just about research institutes, but about bringing all the relevant ecosystems together. We have the university (NUS) nearby, research institutes, and industry, a perfect combination to attract companies to set up R&D centres in Fusionopolis."

Developed by JTC, the Fusionopolis precinct within One-North was designed as a multidisciplinary hub comprising talent from different scientific disciplines and industry sectors. (See Figures 4.1 and 4.2) Its proximity to Biopolis encourages multidisciplinary research collaboration. Construction for Fusionopolis One began in February 2003, and the iconic dual-tower building, designed by Japanese architect Kisho Kurokawa, opened in 2008. It consisted of one block of 22 storeys, the other of 24 storeys, and just over 1.3 million sq ft of space. Continuing the scientifically-themed names started at Biopolis, one tower in Fusionopolis One is named Symbiosis, the other Connexis. Fusionopolis has several unique architectural features. It is the first building in Singapore to use the advanced super-column construction method. This method ensures that the building is not swayed by strong winds, but instead gyrates gently. Another feature is its skybridges that lock the towers firmly and secure the tops and bottoms of the buildings. The towers have 13 sky gardens to help reduce the urban heat island effect while integrating outdoor greenery with the indoors. By 2015, seven new buildings had been added to the Fusionopolis One and Two precincts: Solaris, Nexus, Sandcrawler, Galaxis, Innovis, Kynesis, and Synthesis, bringing the total area to 4.2 million sq ft. There are clean rooms, dry and wet labs, vibration-sensitive facilities. With the completion of Fusionopolis Two, the plan to foster cutting-edge multidisciplinary research and innovation, and catalyse collaborations with industry in one vibrant R&D environment will be realised.

Fig. 4.1.   Fusionopolis.

Fig. 4.2.   The official opening of Fusionopolis by Prime Minister Lee Hsien Loong (second from left) in 2008.

Thus, after 2001, the engineering research institutes took a more multidisciplinary approach to identify areas that they wanted to excel in and set out to achieve capability in those areas that would be comparable to Taiwan's capabilities. After the 1990s, the increasing pace of innovation and shorter product cycles have made R&D capability even more critical to keeping manufacturing in Singapore. Such capability is a more attractive proposition than just the availability of cheap labour, a resource that Singapore can no longer tout.

When the 21st century began, it had become clearer that technologies in different fields were coming together in new and unexpected ways. Keeping the research institutes in silos and working on their own was not exploiting this growing synergy and multidisciplinary approach in science and technology. Prof Chong Tow Chong who was Executive Director of SERC between 2003 and 2010 before becoming Provost of the Singapore University of Technology and Design, recalled in 2015: "When I was Executive Director for DSI in 1998, I didn't have to meet the directors of the other research institutes, like Dr Frans Carpay of SIMTech, Dr Bill Chen of the Institute of Microelectronics. But that was not good. With the formation of SERC and under its leadership, the mindset changed. We began working more closely together. There was collective decision-making to decide what should be the road map. For example, for data storage, what should be the road map? For the chemical industry, what should be the road map? So we came together and the more we came together, the more we can share our expertise and ideas. It made our thinking richer and more relevant. I think the formation of SERC facilitated this and actually led to the concept of Fusionopolis. Without SERC everybody would still be working in a silo. The creation of SERC led to more coordination and we were able to prepare better S&T plans."

The SERC research institutes focus on developing capability in a broad spectrum of technologies: computational methods, manufacturing technologies, materials science, catalysis, nanotechnology, and microelectronics. Singapore's small size becomes an advantage in promoting multidisciplinary research particularly if co-location was possible. Thus came the decision in 2005 to build Fusionopolis as the science and engineering hub to bring the SERC institutes under one roof and attract private sector R&D partnerships. Originally called Techpolis, the hub later changed to Fusionpolis and finally named Fusionopolis with the addition of an 'o' by Philip Yeo because 'Fusionpolis' was harder to pronounce. Prof Chong, then Executive Director of SERC, was tasked by Yeo to bring all the SERC research institutes with the exception of ICES to be domiciled in Fusionopolis at One-North. Phase One was officially opened in 2008 with the Institute of High Performance Computing and the Institute of Infocomms Research moving here. By 2015, with the completion of the other phases, four other SERC research institutes will be in Fusionopolis with the exception of ICES, whose natural home remains with the chemicals industry cluster in Jurong Island.

A*STAR Chairman Lim Chuan Poh explained in the agency's 20th anniversary commemorative publication in 2011: "In the first five-year plan, the way R&D created impact was to create institutes to serve the whole cluster of manufacturing. We were not at a stage to create or look for synergy among the various research institutes to generate a much bigger impact. Later on, such as the biomedical effort, we began building a suite of capabilities for one big industry cluster. This was a fresh approach, a new strategy. This was how the research landscape changed, not in terms of entities, but conceptually." Taking R&D down the multidisciplinary road brought about the Joint Council Office in 2008. (See Chapter 5)

The Institute construct was not the only means that NSTB/A*STAR used to develop capabilities in emerging technologies. Through competitive, merit-based programmes such the Strategic Research Programmes and later the Thematic Strategic Research Programmes many of the technologies transferred to industry today were nucleated. These programmes aimed at large scale R&D

efforts that brought together research institutes, the universities and other public sector organisations. Water research, for example, was funded by NSTB, and the work undertaken by PUB as early as the 1970s was the genesis of NEWater (the PUB's name for its high-grade reclaimed water produced from treated used water that is purified further using advanced membrane technologies, making the water ultra-clean and safe to drink[2]).

Other areas included microelectromechanical systems (MEMS), photonics, biomaterials and logistics, all of which are key mainstream technologies and used pervasively in numerous products. These capabilities were built over the span of a decade, through patient investments, to be ready for industry in the present and future. The capabilities developed were later absorbed by the research institutes as part of their core R&D portfolio and developed further downstream to be amenable for technology transfer. One good example is the Electronics Packaging programme which focused on wafer level packaging, then an emerging technology which anticipated that chips of the future would be highly complex and integrated. The programme was hosted by the Institute of Microelectronics and involved researchers from various local institutes and local and foreign universities. The capabilities spawned later evolved into the state-of-the-art 3D packaging and Through Silicon Via (TSV) technologies. After a decade of R&D investments, Singapore is globally recognised as among the leaders in electronics packaging and it has attracted over $200 million of private sector R&D investments between 2012 and 2014 in this field.

As part of its work of realising value from the huge investments in the RIs, one of the key activities of SERC was to set up mechanisms for transferring technology to the SMEs. This was the GET-Up programme including the T-Up scheme set up in 2002-3 and the establishing of Exploit Technologies Pte Ltd (ETPL). The upgrading of local enterprise was one of the strategies articulated after the 1985 downturn. Working with SPRING Singapore, A*STAR would use GET-Up and T-Up to bring R&D to the larger SMEs. (See Chapter 8)

---

[2] http://www.pub.gov.sg/water/newater/newatertech/Pages/default.aspx.

Chapter 5

The Biomedical Sciences: Research for Better Health

Raj Thampuran and Kong Hwai Loong

Singapore's interest in biotechnology goes way back to the late 1970s at the time that the EDB had succeeded in getting Glaxo (today GlaxoSmithKline) to invest in Singapore and build a plant to produce Zantac. Glaxo was the second pharmaceutical company in Singapore. The first pharmaceutical company to set up in Singapore had been Beechams, a company founded by British chemist Thomas Beecham who, in 1854, started the first-ever factory solely to produce medicines. Beechams had come to Singapore in 1972 to produce the antibiotic Amoxicillin. Beechams' presence in Singapore was an encouragement to Glaxo when it was looking into a place to set up manufacturing of Ranitidine Hydrochloride, the active compound for Zantac, that had been discovered in the 1970s. In 1979, Glaxo inked a deal with EDB to set up a plant in Singapore to manufacture Ranitidine Hydrochloride. The Glaxo plant which went into production in 1982 was the first pharmaceutical plant here to go through a US Food and Drug Administration inspection for which it got a clean bill of health. Zantac would become one of Glaxo's most profitable drugs and would make Glaxo one of the leading pharmaceutical companies by sales revenue. After 1979, local media began carrying stories about biotechnology programmes. The Zantac project would also sow the seeds for Singapore's biotechnology programme. In 1982, Dr Christopher YH Tan, a Singapore-born researcher who was a tenured professor at the University of Calgary noted internationally for his research on Interferon was asked to meet Dr Goh Keng Swee who was at that time First Deputy Prime Minister, Minister for Education, and the chairman of the Monetary Authority of Singapore.

In Singapore to visit his parents, Dr Tan got a call to meet with Dr Goh. It must have been sudden because he recalled in 2015, in an oral history interview, that he turned up before Dr Goh dressed in a 10-year-old Yale T-shirt and shorts, the coolest clothing he had for humid Singapore. Dr Tan recalled that they talked about whether Singapore could have something like Israel's Weizmann Institute, a state-funded institute dedicated to scientific excellence.

Institute of Molecular and Cell Biology (IMCB)

- Inside IMCB is the Advanced Molecular Pathology Laboratory (AMPL), a world-class facility that provides a wide range of pathology-related resources to the biomedical research community throughout Singapore and the region. AMPL is built upon a well-established histopathology laboratory with a reputation for outstanding expertise and knowhow. The laboratory provides capabilities to support basic research, therapeutic target validation and drug safety evaluation.
- In September 2014, Singapore Eye Research Institute (SERI) and IMCB formed a partnership to establish SIPRAD, a joint research programme centred on Retinal Angiogenic Diseases (RAD). The specific aims of the programme were to conduct clinical studies for RAD biomarkers and target discovery, RAD target validation and pathway analysis using *in vitro* and *in vivo* models, and RAD drug repurposing and mechanism of action. SIPRAD's plan to operate as an independent platform aligned to industry standards for generating high quality data has already piqued the interest of several large pharma companies. SIPRAD leverages the combined expertise of SERI and IMCB to create a powerful ally in the hunt for RAD therapeutics. Importantly, the recruitment and longitudinal evaluation of DR and AMD patient cohorts will complement SERI-SNECs growing cohort of well-characterised AMD and DR patients and may facilitate subsequent clinical testing of novel RAD therapies. When successful, SIPRAD has the potential to transform the working relationship between industry and academia and serve as a model for future efforts to translate Singapore's basic and translational clinical science innovations into concrete solutions that address the needs of patients and industry.

He said: "I remember that I wrote out on a piece of paper what would be necessary and gave it to him." Dr Tan recalled in a 2015 oral history interview that on Christmas Eve 1983, he received a phone call from Dr Goh asking him to come to Singapore to take charge of the proposed institute. "Taking charge" meant that he had also to provide building blueprints because Indeco Engineers Pte Ltd, the firm given the task of building the new research institute on the Kent Ridge campus of NUS, had never been asked to build a research institute before. So Dr Tan turned to a friend, Nobel Laureate Dr David Baltimore, the founding director of the Whitehead Institute for Biomedical Research at MIT from 1982 till 1990. Dr Baltimore essentially gave him the building plans for the Whitehead Institute. The original IMCB building on Kent Ridge is the first purpose-built research institute in Singapore. Thus, in

January 1985 when *The Straits Times* ran a story about the setting up of the Institute of Molecular and Cell Biology (IMCB) it was complete with a scale model of the building, a budget and the name of the founding director. It had been one of Dr Goh's last big projects as Deputy Prime Minister. He retired from active politics in 1984 although he would remain active on the boards of several key government agencies into the 1990s. Dr Sydney Brenner, acknowledged as the Father of Biotechnology in Singapore, recalled in a 2015 interview that he remained in touch with Dr Goh even after that.

Interestingly, a few years prior to being asked to take charge of IMCB, Dr Christopher Tan had been offered a chance to start up a biomedical sciences research institute in New York for the Memorial Sloan-Kettering Cancer Centre. One of the men on the Sloan-Kettering interview panel had been Dr Brenner who was London-based and with the Laboratory of Molecular Biology in Cambridge University at that time. Another Singapore researcher had also met Dr Brenner prior to Dr Brenner's involvement with Singapore's biotechnology programme, this time in Kyoto, at a biotechnology conference in 1981. Dr Tan Kok Kheng recalled in 2015: "CSHL (Cold Spring Harbor Laboratory) has a letter from me to Brenner written on 16 December 1981, after I met him at a Kyoto conference. Brenner replied to me on 4 January 1982, saying that he enjoyed meeting me in Kyoto, and hoped to see me when he next goes east again." Dr Brenner came to Singapore for the first time at the end of 1983 when he was the keynote speaker at an ASEAN-EEC seminar on *Biotechnology — The Challenges Ahead*. Interviewed at the opening of the seminar, Dr Brenner was described as "advising the Singapore Government on how to promote biotechnology".[1] He returned in January 1985 as the first Lee Kuan Yew Distinguished Visitor[2] to give a series of four lectures on biotechnology. (During his 10-day stay in Singapore, Dr Brenner was taken on a visit to Dr Tan's mushroom farm and given a bag of shiitake mushrooms which he brought home to share with friends and family!) His lectures were timed with the subsequent announcement of the plan to build the Institute of Molecular and Cell Biology (IMCB) with Dr Brenner as chairman of IMCB's scientific advisory panel and Dr Christopher Tan as the director-designate.

Plans for biotechnology were definitely brewing. In 1990, the National Biotechnology Master Plan was released in which it mapped out five strategies to promote biotechnology as an economic driver in Singapore: technology,

---

[1] http://eresources.nlb.gov.sg/newspapers/Digitised/Article/straitstimes19831115-1.2.41.aspx?q=Dr+Sydney+Brenner&mode=advanced&ct=article&t=straitstimes&page=2&sort=relevance&token=brenner%2csydney%2cdr&sessionid=dbee10b0762e47459aa8e77280fa1012. Retrieved 18 July 2015.
[2] <http://www.lkydvp.sg/dv7.php>. Retrieved 26 June 2015.

manpower, industry and infrastructure development, and raising public aware-ness and interest in biotechnology. Dr Brenner who would be jointly awarded the Nobel Prize in Physiology or Medicine in 2002[3] and honorary Singapore citizenship in 2003, one of the first two such awards,[4] said in 2006 in his accept-ance speech for the Singapore National Science and Technology Medal: "I am very glad there was one person who saw that this was the right approach for Singapore. He was Dr Goh. Dr Goh Keng Swee. He was the architect for what we now see here." And it would be a Dr Goh protégé, Philip Yeo, who would see to the execution of the National Biotechnology Plan in 2001.

The original IMCB building was on the Kent Ridge campus of the National University of Singapore (NUS) on top of a hill next to the National University Hospital and the Medical and Science faculties of this university. The institute would be nurtured by NUS, just like the engineering research institutes would be in 1991. The $25-million budget for the building came from the $60 million budget for developing the biotechnology industry that would be unveiled in the first National Biotechnology Plan in 1990 which was also part of the 1991 National Technology Plan. The 1985 news reported that IMCB was "to do research, particularly into infectious and genetic diseases and cell regulation". The institute would have 12 laboratories, a lecture theatre for 100 people, two teaching laboratories and specialised rooms for various scientific functions. The wet laboratories that are part of biomedical sciences research are more expensive to set up, with protection against accidental infection being neces-sary concerns. Biomedical researchers work with live bacteria and viruses as well as with animals and at some point with humans. IMCB became opera-tional in 1987 but celebrates its founding year as 1985.

Interest in the biomedical sciences is among the earliest of Singapore's interests in R&D for practical reasons. At the end of the 19th century, a unit to handle scientific services had been set up in the municipal government to maintain public health in a crowded port where the comings and goings of ships and people from all parts of the world also brought with them all manner of infectious and contagious diseases. This early concern for public health led to the founding of the School of Medicine in 1905. The School which trained Singapore's pioneer doctors eventually evolved into the Medical Faculty of NUS.

---

[3] Nobel Foundation website www.nobelprize.org: "The Nobel Prize in Physiology or Medicine 2002 was awarded jointly to Sydney Brenner, H. Robert Horvitz and John E. Sulston *"for their discoveries concerning genetic regulation of organ development and programmed cell death""*.

[4] http://eresources.nlb.gov.sg/infopedia/articles/SIP_1871_2012-02-20.html. Retrieved on 18 July 2015. The other was Pasquale Pistorio, CEO of ST Microelectronics who brought the first wafer fab plant to Singapore.

The end-20th century interest in the biomedical sciences was for no less practical reasons. Not long after IMCB became operational, in 1988 EDB formed a National Biotechnology Committee to look into the development of biotechnology as an industry sector for Singapore. EDB attracted a few start-ups to SISIR's Incubator Centre at the Science Park, one of which was Life Technologies which, in 1989, came up with the world's first automatic AIDS diagnostic kit. The EDB set up the Biotech Strategic Business Unit headed by Teoh Yong Sea, who was also General Manager of Singapore Bioinnovations Pte Ltd, Singapore's first biotechnology investment fund. In August 2000, EDB launched the Biomedical Sciences Initiative, a plan masterminded by EDB Chairman Philip Yeo as a major drive to establish BMS as one of the four key pillars of the Singapore economy, alongside Electronics, Engineering and Chemicals. The efforts of BMS were directed at building industrial capital, human capital and intellectual capital. Beyond BMS manufacturing, Singapore aimed to establish a strong base in R&D to attract corporate R&D activities, and to anchor biotech manufacturing activities in Singapore. A budget of $1.48 billion was allocated for Phase 1 of the BMS Initiative.

Up to the end of the 1990s, Singapore's economy was driven mainly by manufacturing in three key sectors: Electronics, Engineering, Chemicals. But particularly in electronics, manufacturing was facing stiffer competition. Yeo who had become Chairman of EDB in 1986 at the time when Singapore was facing its first downturn wanted to set up a new pillar in the economy that would be a little beyond the Third World countries that competed on the basis of cheap labour. He wanted Singapore to move up the economic, knowledge ladder so that whatever it did hopefully would have a better value-add and therefore would translate into better economic returns for the country. The way EDB saw it, biomedical research could be made into an economic pillar through pharma-ceutical production. There were the examples of Beechams and Glaxo to follow. Yeo felt strongly that, of all the biomedical research that Singapore was doing in the 1990s, it was better to channel those efforts into areas with a human aspect rather than dissipate limited resources into plant-based research. In 1995, as part of a national concern for water and food supplies, NSTB had set up an Institute of Molecular Agrobiology (IMA) to spearhead agricultural research at the genetic and molecular levels to establish Singapore as a world-class centre for agrotechnology industries. After Yeo shifted the focus onto the biomedical sci-ences, parts of IMA research that could fit into IMCB were incorporated, while its plant research and collection of plant materials were taken over by a Temasek Holdings company. In 2002, Temasek Life Sciences Laboratory was set up with affiliations with NUS and NTU and with funding from the Temasek Trust Fund.

Announced at the same time as the BMS Initiative was news that Philip Yeo would take over as Executive Chairman of NSTB in February 2001. Recalling the leadership change, Yeo said in A*STAR's commemorative publication: "From 1993 to 2000, I was not involved with NSTB. Then I started to think of trying to promote the biomedical sciences. In 1999, I told Minister George Yeo who was then taking over as Minister for Trade and Industry that we needed to do something about biomedical sciences. I said 'We need to do research. The only biomedical research institute is IMCB.' I said this was not good enough. I volunteered to take over NSTB and restructure it."

The resulting reorganisation led to the creation of two councils to bring together research institutes focused on distinctly different fields of R&D. One was the Science and Engineering Research Council (See Chapter 4, SERC); the other was the Biomedical Research Council (BMRC). Science and engineering research being industry-linked from the beginning had had a head-start on BMS. There were engineering institutes to group into SERC. BMRC only had IMCB which had been working hard on building research reputation through good publications but in a silo. Singapore did not have the spectrum of research capabilities that was needed to help build up the biotechnology sector. Apart from the missing research institutes in the BMS research field, two critical missing essentials were a pool of research and scientific talent and the research infrastructure. Thus, BMRC would focus on establishing research institutes and developing research talent in BMS. At the same time as the announcement of the BMS Initiative in 2000, the decision was made to build Biopolis to house these necessary research institutes as well as co-locate BMS-linked industry players. Biopolis would also be close to the National University Hospital and National University of Singapore in Kent Ridge, the adjoining precinct to One-North's Buona Vista, making a natural research hub for both engineering and biomedical research. At the same time, Yeo also changed NSTB to A*STAR as part of his focus on the development of local scientific and engineering talent, and an "A star" was what all school children in Singapore aimed for. Yeo said in the agency's commemorative publication: "We want to be an agent of change… to be a STAR, to provide the guiding direction for knowledge, change and upgrading in Singapore's economy. A*STAR aspires to be a leading star that guides and inspires our young people, and galvanises our local scientific and engineering community to pursue knowledge for the continued prosperity of Singapore."

Yeo also had another reason for the focus on BMS as a way of developing the pharmaceutical and biotechnology industry. In the end it was all about how to bring better healthcare to people, he said in a 2011 interview for the A*STAR commemorative publication. "So, Chorh Chuan understood that. John Wong understood that. Kong Hwai Loong understood that. George Yeo (the Minister for Trade and Industry 1999–2004) understood that. George was Minister for Health before. They all knew I was not doing science for science's sake. I was doing science for the end point of better healthcare, future treatments for people. That was it. And in the process, I needed to build the biotechnology industry. I needed to build basic research for the industry, for hospitals. And that is the whole value chain. When you make an iPad, you don't take care of people in the hospitals although, if I were a doctor today, I might use an iPad for patients. But every other product has nothing to do with the end point, which is improving the life and health of people. Biotechnology is the only industry that's all the way linked to the biomedical sciences. And it's a very long value chain."

A poignant story lay partly behind Yeo's drive to build up BMS research. As he recounted in *Heart Work*[5]: "Part of my passion for BMS was kindled when things didn't turn out as planned with Tsao Chieh. He was an extremely bright SAF scholar who had joined me in SembCorp in 1995, when I was Chairman from January 1994, as my Special Assistant for Technology. Tsao Chieh was a highly talented person. He obtained three graduate degrees from Stanford University, MSc in Economics, MA in Music and a PhD in Digital Signal Processing in three years! In June 1996 he was diagnosed with liver cancer. His young life suddenly crumbled. He fought hard for his life and his young family. I gave him all the support he needed. There was an experimental drug on clinical trial in the US. The doctors approached him and asked if he wanted to go for it. It might save his life! I got the Permanent Secretary (Ministry of Health) Kwa Soon Bee's approval to allow its immediate importation. He tried it, and we were pinning our hopes on it to pull this young man through. We waited with anticipation. The news came to us: Tsao Chieh had developed an allergy to the drug. I slumped onto my chair upon hearing this news. There I was, watching a young life pass by before me and there was nothing I could do! I felt a personal loss with Tsao Chieh. Singapore, too, lost someone very precious. Our ultimate quest in BMS is to save lives. Some people are misguided when they say that BMS is a risky venture. It is not risk: it is all about caution, and you have to put in time."

Since Yeo is an engineer by training, he roped in three men knowledgeable about the medical sciences. The three were Prof Tan Chorh Chuan, Prof John Eu-Li Wong, and Dr Kong Hwai Loong, all of the Medical Faculty of NUS. (Dr Kong has since gone into private practice.) Yeo has often described the

---

[5] *Heart Work*, page 301.

Identifying Research Areas
Dr Kong Hwai Loong

Medical researchers identify research topics differently than basic science researchers do. Science and engineering people do it differently from biomedical people too. Basic scientists in a place like IMCB would probably identify areas that are virgin territories, because those types of research are more likely to get into top journals. As their end point is primarily high-end publication, they have to always go into new grounds. Being number two is not quite good enough in the research world. So life in the upper echelons of the research world is tough. The good researchers always strive to remain number one in their respective research domains. They constantly look out for uncharted waters. How do they do that? They have to keep abreast of new research findings and trends all over the world, either in their own research domains or in unrelated ones. If a fresh research idea pops up in their heads, they might just seize it and doggedly pursue it, not necessarily knowing what the final end result will be. Medical research is slightly different. It is a bit more utilitarian. Medical scientists ask the question "What do we need to know in order to change the outcome of this disease?" For example, myopia is a big problem in Singapore — what do the medical researchers need to know about this disease in order to prevent the onset or slow down the progression of myopia? With the endpoint already known, they work backwards to get the research to fulfil that end point.

(Continued)

group as "three doctors and a misemployed engineer". Yeo called the group the "Gang of Four", with the word 'gang' to mean a group of people who trust each other, and who have a common vision. As the leader of the Gang, Yeo was most inspirational. Dr Kong has described him as being always full of practical ideas. Two of the doctors, Prof Wong and Dr Kong, are oncologists and Prof Tan is a nephrologist. All three were NUS graduates and by the time they become members of the Gang of Four, had considerable experience not only in research but also organising research and administration. They were brought together to essentially answer the question: How do you actually translate research into medicine or treatment? Dr Kong had spent two years until 1997 in Cornell Medical Centre, New York Hospital, basically involved in a research area called anti-angiogenesis in which he researched the use of drugs to stop blood vessels from growing. He recalled in a 2014 oral history interview that he was probably one of the first oncologists to be sent for a research stint in addition to clinical training because his head of department, Prof John Wong, felt

(*Continued*)

Basic science is always working forward. Basic scientists go into unchartered waters. Sometimes they have only the vaguest clue what will happen at the end of the journey. But that is the fun part about research as well. Medical researchers are more pragmatic in their research approaches. It is therefore important for BMRC to understand the different starting points and aspirations of the basic scientists vis-a-vis the medical researchers. When panel advisers review the funding applications for basic research, the scoring criteria would be slightly different from those for medical research. One starts at the beginning, while the other starts at the end. And therefore it is interesting if the two camps can work together, and when they are able to meet in the middle, magic can happen! BMRC hopes to marry these two research communities because they see biomedical science from two different, yet complementary, points of view. The basic scientists have a very wide-angle lens. They are quite free to go in any research direction. The likelihood of success, as defined by commercial output, may be relatively low because there will be many hits and misses with such a wide arc of fire. The medical researchers, on the other hand, tend to be more narrowly focused on their chosen areas of research. Medical researchers can help basic scientists to identify specific areas of research that are more likely to bear fruits in the healthcare market. In a relatively small country like Singapore, it would be a distinct competitive advantage if we can successfully integrate basic and medical research to address unmet healthcare needs.

that it was time for oncology in Singapore to take on not just a clinical flavour but also to have some research perspective, and so Dr Kong was sent to spearhead this area of medicine. The awards that Dr Kong garnered at this time brought him to the attention of EDB Chairman Philip Yeo and a role in setting up the next BMS research institute after IMCB.

While NSTB was doing good engineering research, in the 1990s there was no focused research geared towards human health outcomes apart from the limited research in NUS. (See Chapter 6.) IMCB was focusing on good explorative basic research at that time and building a solid reputation for the quality and high standard of its research work. The papers of its researchers were getting published and cited in learned journals such as *Cell, Science,* and *Nature.* It was laying a solid foundation that would ease the transition into the biomedical research era. Without IMCB, it would have taken much longer to recruit scientific talent and attract the necessary critical group of top scientists to jumpstart the BMS Initiative. IMCB played a critical role in being the core

Biopolis

One-North, where Biopolis and Fusionopolis are located, was the brainchild of Philip Yeo, Chairman of NSTB (before it became A*STAR) and Lim Neo Chian, JTC Chief Executive Officer. Their bold vision was to establish the Biomedical Sciences as a key pillar of the Singapore economy, and Biopolis was to be the iconic space that would show the scale of the ambition for Singapore to be Asia's biomedical sciences hub.

The name "Biopolis" was coined by Nobel Laureate Sydney Brenner, along with Yeo, a self-professed ancient Greek history addict. "Polis" literally means "city" in Greek. Yeo recalled: "I took charge of the then NSTB on 1 February 2001. Brought Dr Sydney Brenner to the spot where Biopolis is today. I told Sydney that we (JTC and NSTB) will build a Home for Biomedical Research." Ground was broken for Biopolis on 6 December 2001. Designed by award-winning British-Iraqi architect Zaha Hadid, driven by Yeo and Lim, and with construction going on virtually 24/7, Phase 1 was completed in 2003 with a two million sq ft, seven-building integrated complex linked by sky bridges. Five research institutes — GIS, BII, BTI, IMCB, and IBN moved in by 2004. (See Figures 5.1 and 5.2)

Fig. 5.1.   Biopolis at One-North.

(*Continued*)

(*Continued*)

The Biopolis cluster later expanded from the original seven buildings through five phases of construction to 13 buildings, with a total floor area of 3.7 million sq-ft. Phase 1 was spearheaded by JTC and the remaining phases were developed by private developers. The buildings in Biopolis are aptly named— Phase 1 saw the creation of Genome, Centros, Matrix, Nanos, Proteos, Chromos and Helios; Neuros and Immunos were developed in Phase 2; Phase 3 comprised Synapse and Amnios; Phase 4 saw the development of Procter & Gamble's Singapore Innovation Centre; and Phase 5, Nucleos, was completed in 2014. P&G's Singapore Innovation Centre can house 500 researchers and is the second of only two such centres in Asia. Today, Biopolis is a biomedical sciences research hub that proudly hosts more than 50 biomedical companies and 10 A*STAR's research institutes and consortia with a total working population of 5,600 people consisting of researchers and non-researchers. A*STAR alone employs close to 1,800 researchers including Research Scientists and Engineers (RSEs) and Research and Technical Support (RTSs) from 53 different countries.

Fig. 5.2.   Opening of Biopolis Phase II.

around which the crystal grows. What Yeo wanted was for the research to produce some tangible applications to human health. If a commercial outcome came along, that was a bonus but the research should at least advance medicine, advance health outcomes. Although Yeo did not use the term, what he wanted to establish at A*STAR was translational research. While the phrase "translational medicine", or "translational research" is commonly bandied about, very few research institutes actually execute that translation. What Yeo did with Biopolis and the BMRC institutes established after 2000 was to bridge the gap between research and translational medicine.

The BMS Initiative faced four main challenges at the beginning of the effort. The first was the question of laboratory space, the second the absence of research institutes essential to the Initiative. Linked to this was getting the necessary high-level research manpower to jumpstart the institutes. The final challenge was how to get researchers to work outside silos. The first challenge was met with the construction of Biopolis. It is a physical manifestation of Singapore's ambition and aspiration. Said Yeo in the Fullerton-SJI Leadership lectures in May 2010 on the construction of Biopolis and what it achieved: "Eighteen months, seven buildings, 2 million square feet".[6] Before Biopolis, there were lab spaces across the island, in different parts of the island. But in order for the world to recognise Singapore as a major biomedical hub, it is quite important, Yeo felt, to have a physical structure to not only become a lighthouse for people to take note of us, but also to be a lighthouse for the scientists who are on the ground to know that there is one mega lab somewhere out there in Biopolis which has resources that all can use and to seek collaboration. So the formation of a mega integrated lab which is Biopolis with not only adequate resources but also internationally prominent ones was one challenge.

The second challenge was to set up the requisite research institutes to establish a spectrum of capabilities and to get them to spread the knowledge gained downstream so that their work would not just end with the publication of papers but would actually percolate down the value chain and translate into a product or application with impact on healthcare. It could be a diagnostic tool or a therapeutic product which, of course, would be the Holy Grail. For instance a diagnostic tool came out in the SARS crisis in 2003. The SARS crisis underlined the importance of a biomedical research outfit like the Genome Institute of Singapore and cleared any doubts about basic research because in the shortest time possible the researchers there sequenced the genome of the

---

[6] <https://www.youtube.com/watch?v=HIlthlw4ITs>

SARS virus. They got the diagnostic kit out to the hospital labs to screen patients and so the health services were quickly able to separate SARS from non-SARS and unblock the freeze on economic activities. The kit demonstrated the route from basic research to medical application or translation and made supporters out of doubters.

Even before Biopolis was ready, the research institutes that would establish a spectrum of capabilities were being formed. After Dr Kong was roped into the Gang of Four, in 1999 Yeo asked him to look into how to set up a DNA-based research organisation. With the help of two molecular scientists, Dr Kong cleared the hurdles that setting up such a research programme in Singapore would entail and in 2000, the Singapore Genomics Programme (SGP) was set up. In 2001 it become the Genome Institute of Singapore (GIS) with Dr Edison Liu as the founding executive director. At that time, the Human Genome Project was still rapidly unfolding. Started in 1990, the full sequencing of the human genome was completed only in April 2003. Gene sequencing has become one of the key technologies for tackling disease. GIS, in pursuing

Genome Institute of Singapore (GIS)

- In 2003, during the SARS crisis, a team of researchers at the Genome Institute of Singapore (GIS) collaborated with Roche Diagnostics to co-develop a SARS detection kit. The kit used Roche's Polymerase Chain Reaction (PCR)-based platform technology to rapidly and accurately detect a SARS coronavirus genetic sequence which was sequenced at GIS. The team then tested the kit clinically in the Singapore General Hospital (SGH).
- GIS and SingHealth have entered a research collaboration titled Personalised OMIC Lattice for Advanced Research and Improving Stratification (POLARIS). POLARIS is an initiative within the scope of stratified medicine, which aims to deliver better patient outcomes through research. With POLARIS, clinical researchers from various SingHealth institutions such as Singapore General Hospital, National Cancer Centre Singapore and Singapore National Eye Centre can identify critical biomarkers in patients, resulting in a more targeted approach to treating diseases. POLARIS utilises Next Generation Sequencing (NGS) technology in the laboratories to conduct tests. NGS is a fundamentally different approach to sequencing and has the ability to trigger ground-breaking discoveries, igniting a revolution in genomic science. POLARIS is currently developing a clinical assay for corneal dystrophy. Gastrointestinal and lung cancer panels are also in development stages. POLARIS is working closely with clinicians and scientists at both SingHealth and GIS campuses.

The SARS Crisis

SARS (Severe Acute Respiratory Syndrome) broke out in Singapore in late February 2003 with the return of three young women from a holiday in Hong Kong. They fell ill and were hospitalised for what looked like pneumonia. While in a hotel there they had picked up a new virus now known as the SARS coronavirus (SARS-CoV) from a fellow guest, a doctor who had been infected in Guangzhou, China. Two of the women recovered but the third set off the SARS crisis with a string of transmission cases that affected healthcare workers, family members and friends. SARS struck fear in the country and contracted the economy by 4.2% in the April-June quarter year-on-year.[7] A total of 238 people were infected, of whom 33 died. Forty-one per cent of those infected were health care workers who picked up the virus when people were hospitalised with pneumonia-like symptoms. By then this mysterious pneumonia-like illness had surfaced in some 20 countries. It was a global epidemic and controlling the spread of this highly communicable disease became paramount. Identification of the virus and isolation of the infected became critical to containment of the spread. The Genome Institute of Singapore (GIS) would play a vital role in this containment and highlight the fact that research institutes can contribute to the economy in unexpected ways.

Once it became clear that this was an unknown virus GIS executive director Dr Edison Liu went to work. Speaking in an oral history interview in 2015 he recalled: "What we did at GIS was we mobilised the entire institute towards solving the SARS problem. I knew for a fact that the SARS solution was a genomic solution. You had to sequence the virus in order to get at the diagnostics, to get at everything downstream to it. So we had a phenomenal spirit that came around. At that time we had about 110 people in the institute. I remember the first meeting. I called everybody and said 'Listen we have a national crisis. I am calling for volunteers, I am calling for this institute to drop everything they are doing to work on this problem. Not everybody's skillset is going to be appropriate for this but I want everybody, no matter what they are doing to help out in the best way they can. So we fractioned it out to those who were going to sequence, those people doing the cell biology, those people putting together the kit that was going to come out.

(Continued)

---

[7] http://eresources.nlb.gov.sg/infopedia/articles/SIP_1529_2009-06-03.html?v=1&utm_expid=85360850-6.qNOOYF40RhKK6gXsQEaAJA.1&utm_referrer=https%3A%2F%2Fwww.google.com.sg%2F. Retrieved on 16 May 2015.

(Continued)

In the morning we met with the team leaders and said 'OK, what are you going to do today?', and so forth and so on. And at the end of the day, we met with the team leaders and said 'what happened today?' We did a website where everybody could check up on each other and part of my responsibility was to work with the different ministries, the different units outside, the different hospitals while my colleagues were the ones doing the basic work. Our administration would support everybody and they did a fantastic job. Because they had to have temperature taken, they had to register everything. The sequencing team worked 24/7. I mean, they slept in the hallway. In those days sequencing was not that easy. We missed being the first by probably 36 hours before the Canadians put it in the web."

Said Dr Liu: "Once we knew that the virus was non-mutating we styled primers for it to make diagnostics. We had a team that put together the diagnostic kit with instructions on how to make more. We delivered it to the various hospitals, Ministry of Health, National Environment Agency and so forth. We gave them the diagnostics so that they could track the virus in culture. And we started drugs for it. We found that the drug that was being used in Hong Kong was totally worthless but that beta-interferon seemed to work, and lucky for us at that time, the whole epidemic went away. But what that experience did was it made obvious to all the decision-makers that their investment in the biomedical sciences was not just to get cash return in the stock market. It was for national security. And here's the thing. Singapore was cited by WHO as an exemplary country in its response to the SARS crisis. China was considered a problem. Singapore beat China in resolving the SARS crisis.

"It took a total of three months from when the teams were organised to work on the problem to getting the diagnostic kit. Actually the sequencing and the analysis of it took about a month. … What working on SARS did was it really proved internally that working as teams was really powerful. It also proved to A*STAR that their investment in science was extremely valuable. A*STAR's institutes became part of the pandemic response in MOH thereafter. So when H1N1 came around we were there immediately to respond to it. It was so cool. Completely different from the SARS crisis. Now it's all embedded."

The last SARS case was isolated in early May and Singapore was removed from the World Health Organisation's list of SARS-affected areas on 31 May. Singapore's experience became a model of how to respond to a potential pandemic.[8]

---

[8] SARS: how a global epidemic was stopped. World Health Organisation, c2006.

Bioinformatics Institute (BII)

Since 2007, BII's Genome and Gene Expression Data Analysis Division has identified and validated biomarkers for risk assessment, early detection, differential diagnostics, tumour classification, prognosis, prediction and therapeutic targeting of breast, lung, ovarian carcinoma (OC) and several other cancers. Among these, they made breakthroughs in ovarian cancer research:

- Identified two poorly prognosed OC subtypes associated with CHEK2 germline mutation and non-CHEK2 somatic mutation gene signatures;
- Discovered the key role of MECOM/EVI1 complex locus amplifications and EVI1 oncogene pathway gene expression in high-grade OC initiation and progression, as well as the clinical significance of these findings for early and differential diagnostics, survival risk prognosis and disease outcome stratification of the patients; and
- Developed a multi-gene microRNA/mRNA–based disease classifiers stratifying and personalising prediction of patient's risk of cancer recurrence and sensitivity to chemotherapy following surgery treatment.

BII together with the Singapore Eye Research Institute (SERI) have developed novel antibacterial agents that work extremely well against resistant bacteria, a major healthcare issue worldwide. A highly significant discovery is that the molecules appear to stop bacteria from developing resistance. This is a major milestone and has led to the establishment of a spin-off company called SinsaLabs to develop these molecules as therapeutics.

- Triggered by the H1N1 swine flu in 2009, a BII research group developed the FluSurver research tool. The BII FluSurver is designed to help clinician-scientists and researchers rapidly screen patient-specific influenza genome sequences for mutations that potentially induce drug resistance, antigenic shift, or other clinically relevant features. The FluSurver is now used as part of the World Health Organisation (WHO) surveillance network. Tools like the BII FluSurver have enabled WHO to react more quickly to the recent virus outbreaks such as H7N9, Ebola and Middle East Respiratory Syndrome (MERS).

genomic research was integrating medicine, technology and biology, and in 2003 during the SARS crisis, it would come into its own.

To have the necessary range of research capabilities, a series of research institutes and consortia were set up after 2000. In the same year as the founding of GIS, the Bioinformatics Institute (BII) was founded by Dr Gunaretnam

Rajagopal as an IT services and bioinformatics support unit. Gene sequencing requires specialised high-power computing. BII became a biological research institute in 2007 with Dr Frank Eisenhaber as its director with the goal of computational-biology-driven life science research into biomolecular mechanisms. The institute develops appropriate computer-based theoretical research tools and collaborates with experimental and clinical groups from academia as well as with industry.

The Bioprocessing Technology Institute (BTI) started life in 1990 as the Bioprocessing Technology Unit in NUS with EDB funding. The unit facilitated research in biotechnology and the process development needs of the biotechnology industry. In 1994 it was taken over by NSTB and renamed Bioprocessing Technology Centre, and in 2003 it became the Bioprocessing Technology Institute with Prof Miranda Yap as its founding executive director. This institute uses its core expertise in expression engineering, animal cell technology, stem cells, microbial fermentation, product characterisation, downstream processing, purification and stability with supporting proteomics and microarray platform technologies. It bridges the gap between discovery, process

Bioprocessing Technology Institute (BTI)

Biologics manufacturing, a knowledge-intensive and high value-added activity, is a prime example of how Singapore's efforts to build up biomedical sciences R&D capabilities have successfully attracted large-scale manufacturing investments. For over two decades, BTI's mission has been to develop manpower capabilities, spearhead research in bioprocessing, and to strengthen linkages between the laboratory and industry. BTI has been instrumental in enabling Singapore to gain a foothold in the world's rapidly growing biologics sector. To support manpower development for the biologics industry in Singapore, BTI has been running the Bioprocess Internship Programme (BIP) since 2005 to train skilled talent in biologics and process development for industry. To date, BTI has trained about 150 people in this programme, of which almost half went on to join industry. BTI has also spun off about 70 of its staff to industry since 2003. Having a strong local talent pool enables companies to ramp up their operations quickly and is a key value proposition for companies to invest in Singapore. Through BTI, Singapore has attracted commercial-scale biologics plants from major biologics players — GSK, Lonza, Roche, Abbvie, Baxter Bioscience, Novartis and Amgen — that together employ more than 1,700 people and brought in $2.8 billion in investments since the first biologics manufacturing investment in 2007, thus enhancing Singapore's status as a global hub.

A Biomedical Sciences Pioneer: Prof Miranda Yap (1948–2015)
Prof Lam Kong Peng, Executive Director, Bioprocessing Technology Institute

Born in August 1948, Prof Miranda Yap passed away in October 2015. Founder-Director of A*STAR's Bioprocessing Technology Institute (BTI), Prof Miranda Yap was one of the pioneers in the Biomedical Sciences (BMS) Initiative. Her efforts went a long way to develop the biologics industry sector in Singapore. She introduced mammalian cell culture technology to Singapore that was critical for the scale-up production of biologics. Today, BTI is the nation's flagship institute that has helped to attract over $2.7 billion in investments to Singapore in the past eight years, and the institute anchors multinational biologics manufacturing facilities that together employ over 2,000 workers. In the words of Philip Yeo, former Economic Development Board Chairman and the A*STAR Chairman who got the Biomedical Sciences Initiative going: "Miranda was a great asset to EDB's BMS industrial effort. She built A*STAR's Bioprocessing Technology Institute. Hence EDB attracted eight biologics companies to Singapore, supplying the world with the latest drugs for cancer and other diseases."

BTI was founded as the Bioprocessing Technology Unit (BTU) in 1990 at the National University of Singapore (NUS) Chemical Engineering Department with a $6-million grant from the Economic Development Board (EDB). As founding executive director of BTI, Prof Yap nurtured the institute until her retirement in 2011. Her leadership role in the establishment of the Consortium for CHO (Chinese Hamster Ovary) Cell Genomics with the University of Minnesota and the US Society for Biological Engineering, and with the participation of virtually all the large multinational biopharmaceutical companies, firmly placed BTI as one of the premier institutes for bioprocess science and engineering. These efforts together helped to position Singapore as an attractive destination for biologics research, development and production.

Prof Yap's other notable achievements include starting the Center for Natural Product Research (CNPR) in 1993, which was a joint venture between Glaxo Wellcome and the Institute of Molecular and Cell Biology (IMCB). This was later spun out as MerLion Pharmaceuticals in 2002. She also set up Genset Singapore

(*Continued*)

development and commercialisation. In 1997, it set up incubator units to nurture SME biotech companies in Singapore.

Also formed in 2003, and the world's first, was the Institute of Bioengineering and Nanotechnology (IBN) to focus on research at the interface of these two multidisciplinary and highly diversified fields that brings together science,

(Continued)

Biotechnology Pte Ltd, a joint venture company with Genset (France) that focused on oligos synthesis in 1998. In 1999, Prof Yap established the Biopharmaceutical Manufacturing Technology Centre (BMTC), a state-of-the-art GMP facility for the production of protein therapeutics and this was spun out from BTI as A-Bio Pharma Pte Ltd in 2003. Prof Yap was also a co-founder of VeriStem Technologies in 2009, a company focused on developing stem cell therapies. These spin-offs contributed to the vibrancy of the biotechnology start-ups and spin-offs scene in Singapore.

Aside from R&D endeavours, Prof Yap was also passionate about developing and nurturing scientific talent. She was very active in the graduation of numerous PhD and Masters' students in biotechnology and biochemical engineering, many of whom are now senior scientists and managers in academia and companies. From 2006 to 2009, she was concurrently appointed executive director of the A*STAR Graduate Academy (A*GA) and was in charge of scholarship development at the agency. In her illustrious career, Prof Yap was most noted for setting up in 2006 the Bioprocess Internship Programme (BIP) to prepare science and engineering graduates for entry into the biopharmaceutical industry, thus supporting the growing manpower needs of the industry.

Prof Yap's outstanding achievements in education, R&D, and scientific management were recognised by numerous awards and accolades. She was elected as foreign associate into the prestigious US National Academy of Engineering in 2006, becoming the first foreign female associate and the only Singaporean engineer to be inducted into the academy. In 2009, she was given the Asia-Pacific Biochemical Engineering Conference Award, and also the Distinguished Service Award by the US Society for Biological Engineering. In the same year, she was also awarded the inaugural President's Science and Technology Medal in recognition of her distinguished and sustained contribution to Singapore's development as a global R&D centre.

As a R&D manager, Prof Yap built up an excellent work culture at BTI where there was also fun along with the hard work. She will be remembered every time durian season comes around and when BTI organises its annual durian get-together. Prof Yap was very partial to this pungent fruit and managed to cultivate a taste for it in the multinational A*STAR research community.

engineering and medicine. Its founding executive director is Prof Jackie Ying. The institute's goal is to create a critical knowledge base in bioengineering and nanotechnology, and to make significant impact in healthcare through the generation of new materials, devices, instruments, systems and processes. It has tremendous potential for improving health and quality of life in areas such as

Institute of Bioengineering and Nanotechnology (IBN)

- IBN has contributed significantly to the development of Singapore's local entrepreneurial ecosystem over the last 10 years through the establishment of eight spin-off companies: Curiox Biosystems, SG Molecular Diagnostics, SG Microlab Devices, HistoIndex, CellSievo, Baldr Biosystems, Invitrocue and Wet Alert. In the next 10 years, IBN hopes to contribute further to the next phase of Singapore's economic development by nurturing a vibrant environment to support local spin-offs and start-ups, and working even more closely with the clinical community to translate new technologies for the benefit of healthcare.
- Delta Electronics Int'l (Singapore) signed an MOU with IBN in April 2014 to establish a research centre in IBN's premises at Biopolis. This represents a projected cumulative research investment by Delta of at least $20 million over six years. Under the MOU, Delta and IBN will conduct joint R&D in Biosensors, Bioassays, Diagnostics Devices and Systems.

drug and gene delivery, cell and tissue engineering, biodevices and diagnostics, pharmaceuticals synthesis, and green chemistry.

A year later, in 2004, came the Centre for Molecular Medicine which in 2007 became the Institute of Medical Biology (IMB) with Prof Birgitte Lane as its founding executive director. It focuses on issues at the interface between basic science and medicine to facilitate the development of translational research. Its role is to work closely with scientists and clinicians to support, inform and refine each other's strengths and specialisations to improve the translational process. This institute's researchers were early founders of the Singapore Stem Cell Club (now Singapore Stem Cell Society), the Singapore Skin Club, and BMRC's first Postdoctoral Society. These networking setups foster dialogue between clinicians and scientists here. Around 2005, the Singapore Stem Cell Consortium was also set up, with Prof Roger Pederson and Prof Lee Eng Hin as the Co-Chairs to facilitate stem cell research in Singapore. The Singapore Stem Cell Consortium was subsequently absorbed into the Institute of Medical Biology. The Singapore Bioimaging Consortium (SBIC) was also set up around 2004 to provide imaging capabilities for research at Biopolis. It was founded by Sir George Radda.[9] It is one of the few research centres in the world focused on clinical imaging.

---

[9] In 2015 Sir George Radda was conferred the Honorary Citizen Award for playing a pivotal role in Singapore's biomedical sciences industry.

Institute of Molecular Biology (IMB)

- The Skin Research Institute of Singapore (SRIS) is a collaboration between IMB, the National Skin Centre (NSC), and Nanyang Technological University (NTU). It was announced in September 2013. This collaboration will harness the expertise of scientists, clinicians and engineers and the unique phenotypes in Singapore to foster high impact, inter-disciplinary skin research designed to translate into improved health outcomes and quality of life. This $100-million initiative aims to create a new skin research institute to support world-class skin research and collaborations with industry.

- IMB has a research group that leads a strategic programme on Genetic Orphan Diseases to study rare genetic diseases and identify the mutations that lead to them. This is to help reveal the biological processes behind more common diseases. For example, findings from a 2009 study that focused on a rare disease causing premature ageing improved understanding of the ageing process and led to a collaboration with AmorePacific, South Korea's top cosmetic firm. More recently, the Genetic Orphan Diseases programme in collaboration with Kandang Kerbau Women's and Children's Hospital helped to diagnose a baby girl with a rare disorder which led to her successful bone marrow transplant and near complete recovery.

In 2007, the Singapore Institute for Clinical Sciences (SICS) was set up to accelerate the translation of basic discoveries into new diagnostics and therapeutics, with Prof Judith Swain as its founding executive director. It focuses on research on clinical applications, the use of innovative approaches and technologies that enable the study of human health and diseases, especially infectious diseases, metabolic diseases and cognitive development. This institute is located in the Brenner Centre for Molecular Medicine within NUS and next to NUH. It collaborates with the public hospitals, Singapore Bioimaging Consortium, Bioprocessing Technology Institute, Institute of Molecular and Cell Biology, and the Singapore Immunology Network.

Experimental Therapeutics Centre (ETC) was founded by Prof Sir David Lane in 2006 with the mission of guiding early-state drug discoveries towards proof-of-concept in man. It bridges the gap between basic and clinical research. In 2008, the Singapore Immunology Network (SIgN) was set up to grow immunology capabilities to support efficient translation into clinical applications in areas of medical needs. Its founding chairman is Prof Philippe Kourilsky. SIgN researchers focus on human immunity during infections and inflammatory conditions including cancer by using human tissues to

Singapore Bioimaging Consortium (SBIC)

- In October 2014 SBIC partnered with Bruker Corporation, a leading manufacturer of scientific instruments for molecular research, to open Bruker's first preclinical imaging centre in Asia, and the second such centre worldwide. Located in SBIC, the new SBIC-Bruker PCI centre will provide demonstration, training and applications support for Bruker's preclinical portfolio and multi-modal imaging platform. Through the new SBIC-Bruker PCI Centre, A*STAR's research institutions will be able to leverage on Bruker's advanced imaging platforms to boost their research efforts, while Bruker can tap on SBIC's talent and imaging capabilities. Besides gaining access to SBIC's talent and imaging expertise, the SBIC-Bruker PCI Centre will allow Bruker to foster closer collaborations with SBIC's partners such as those in the clinical and academic community, as well as the pharmaceutical companies with research operations in Singapore.

- The SBIC-Nikon Imaging Centre (NIC) is a joint core imaging facility with state-of-the-art equipment for light microscopy developed in partnership between SBIC and Nikon Singapore Pte Ltd. Established in 2007 at SBIC, SBIC-NIC also has several corporate contributors including Zugo Photonics, Chroma Technology, microLAMDA Pte. Ltd., Tokai Hit, Okolab, Prior Scientific, Newport, Einst, Photometrics, Andor Technology, Laser 21 and Coherent Inc. SBIC-NIC aims to:

  — Promote innovation in biological research by providing access to cutting edge microscopy and imaging equipment;

  — Provide training courses on basic and advanced light microscopy techniques for the benefit of Biopolis and the regional research community;

  — Introduce the latest state-of-the-art light microscopy and imaging to the research community;

  — Serve as a learning platform for our regional corporate partners and contributors;

  — Develop new microscope set-ups and imaging techniques in response to feedback from the users of NIC@SBIC.

complement animal models in an effort to translate bench discoveries into medical treatments. SIgN was also designed to attract R&D investments from industrial companies in order to create economic impact. The network's researchers work with over 12 pharmaceutical and biotech companies to co-develop new products and novel medical treatments. More of such

Singapore Institute for Clinical Sciences (SICS)

- In 2013 A*STAR and NUS jointly established the $148-million Singapore Centre for Nutritional Sciences, Metabolic Diseases and Human Development (SiNMeD). It is an example of the union between scientific research, translational initiatives and clinical practice. This collaboration between the NUS Yong Loo Lin School of Medicine and SICS is set to become the leading centre in Asia for research in the nexus between nutritional sciences, metabolic diseases and human development. SiNMeD will focus on fundamental, clinical and translational research to understand the role of nutrition and early development in the onset and progression of obesity and metabolic diseases like diabetes in the Asian context. SiNMeD's research programmes will expand on the success of the unique and internationally recognised GUSTO (Growing Up in Singapore Towards Healthy Outcomes) birth cohort study. The GUSTO cohort study is a collaboration between SICS, Kandang Kerbau Women's and Children's Hospital, and National University Hospital. It is one of the most intensively studied cohorts in Asia of mothers and children, growing in strength in epigenetic analysis, and involving over 100 investigators in Singapore and international collaborators in the UK, New Zealand and Canada.
- SiNMeD also features the joint SICS-NUHS Clinical Nutrition Research Centre (CNRC), the most advanced center for nutrition research in Asia with a full range of nutritional, metabolic and energetics, and sensory infrastructure including the only whole body calorimeter in Asia.

Experimental Therapeutics Centre (ETC)

- ETC and Drug Discovery Unit (D3) have developed a new cancer drug called ETC-159, together with Duke-NUS Graduate Medical School. This compound homes in on proteins that cause excessive cell growth when they go rogue, leading to cancers. ETC-159 is currently being tested on patients at the National Cancer Centre Singapore and the National University Hospital. Clinical trials will be extended to the United States later. If successful, ETC-159 could emerge as Singapore's first cancer blockbuster. This is also the first time that a publicly-funded drug candidate, discovered and developed in Singapore, is being tested on people here.

Singapore Immunology Network (SIgN)

A SIgN team of scientists have discovered a new strategy to prevent the dengue virus from escaping the host immune system. This exciting discovery might provide a clue to the development of a vaccine that can give full protection from all four serotypes of the dengue virus. This new strategy deploys a Trojan horse tactic by introducing a genetic mutation of the MTase, an enzyme found in dengue virus that facilitates chemical modification of its genetic material to escape detection in the host. Once the initial cells are infected by the weakened MTase mutant viruses, the body recognises them as foreign, thereby triggering immunisation responses.

networks and consortia to grow industry collaboration would be established. (See Chapter 7) Essentially, between 2000 and 2010, institutes were established that would have core research strengths in Bioprocessing, Bioimaging, Genomics and Proteomics, Molecular and Cell Biology, Bioengineering and Nanotechnology, Computational Biology, Immunology, Drugs and Biologics, and Discovery and Development.

Whales and Guppies

The third challenge that the BMS Initiative faced and which was linked to the second challenge was the lack of credible research manpower. For the R&D enterprise to be viable and sustainable over the long term, there had to be a critical pool of Research Scientists and Engineers (RSEs) made up of senior RSEs and a pipeline of local RSEs. While the National Science Scholarships would attract young Singaporeans to work towards their PhDs and to eventually form the essential core of the R&D ecosystem, there was a lack of senior RSEs especially in the BMS area. Faced with the urgency of getting the BMS Initiative going and to jump-start it, Yeo embarked on a strategy of what he has termed "kidnapping" top scientific talent which he called "whales". These "whales" would nurture the "guppies" or young Singaporean scientific talent to bring them up to speed and to eventually take charge.

Together with Dr Brenner who had been advising Yeo on how to get the Biotechnology Plan going, Yeo had a list of 100 scientists whom he wanted to try to attract to Singapore. One of Yeo's tactics was to get talent to come to see what Singapore was trying to do and get them interested in becoming part of it. (Box: Why I Came, page 126) The first top scientist that he attracted to

Singapore to be Genome Institute of Singapore's founding executive director was Dr Edison Liu who was at that time the scientific director of the National Cancer Institute's Division of Clinical Sciences in Bethesda, Maryland. Next to arrive was Prof Sir David Lane and his spouse, Prof Birgitte Lane, both from the University of Dundee. Sir David is noted as the discoverer of p53, a protein involved in at least 50 per cent of cancers, while Prof Birgitte Lane is known internationally for her work on keratins and her involvement with A*STAR began with the then Centre for Molecular Medicine (CMM) where she was a Programme Director. In 2007, she became the founding executive director of IMB. When Dr Christopher YH Tan left in 2004, Sir David took over the reins of IMCB as its Executive Director till 2007, becoming Chairman of BMRC, and founding Experimental Therapeutics Centre (ETC) in 2006, before taking on the role of Chief Scientist in 2009.

In 2003, Prof Jackie Ying who was one of the youngest tenured professors at MIT took up Yeo's invitation to come to Singapore and became founding executive director of the Institute of Bioengineering and Nanotechnology. Yeo said in a 2007 interview as quoted in the A*STAR commemorative book: "These scientists of international stature are role models for the younger generation of Singapore scientists. They have helped Singapore to catch up with established R&D centres." Today Singapore's R&D ecosystem has more than 60 nationalities in it.

The numbers may increase or decrease as personal and career needs dictate. Today's international scientific talent is highly mobile including Singaporean talent, and the departure of high-profile talent is sometimes viewed as failures of the system. This is not so. While it is logical to think that the longer such talent stays, the better it is for us, this is rarely the case. What matters is whether they leave behind a good system and good teams and whether they have stayed long enough to do that. And the pioneering whales have done all that. For example, Dr Edison Liu stayed for about 10 years and established GIS to be one of the top institutes, created the SARS kit, recruited many post-docs, and nurtured a stream of PhDs. He left behind good legacies and he was even appointed Chairman of the Health Sciences Authority for a while. So these top people left behind positive things. Some have remained and kept up their association with Singapore's R&D ecosystem in various capacities, be it as advisors on the scientific boards or councils or even as researchers. Another plus in the mobility of scientific talent is that those who leave Singapore as part of career advancement or for personal reasons are rarely permanently lost. Many become links in the international scientific network and continue to play roles to enhance Singapore's R&D credibility. In the early years what the presence of these top

Why I Came
Dr Edison Liu, Founding Executive Director of Genome Institute
of Singapore

No. 1, I was really intrigued by the idea that science could make a difference for
a whole nation. That was what was enunciated, that they wanted to make the
biomedical sciences as a pillar of the economy, and I was very fascinated about
how that could happen. That was a huge part of it. The second reason. When I
came in November 2000 I wanted to meet as many decision-makers as I could.
I've been around in the world. I have seen how King So and So wanted to have
an institute in his name and that's the only reason that institute is there. I didn't
want to be part of that. I wanted to be part of something that would make a
difference on a national scale, and in Singapore this was clear. From the Deputy
Prime Minister, at the time Lee Hsien Loong, and to all the ministers, to all the
academics, this was a national enterprise. So whatever I did was beyond science.
It was going to be important to a nation, a country. That was a great challenge.
The third was that it was a real opportunity with a lot of resources to be able to
craft where genomics was going to be, not where it is now. Because what Philip
Yeo did was invest in a guy with an idea of where it should go but has never been
proven. And that was really very exciting. It was a gamble but it was a gamble that
with the resources that I had, I had a real chance of making a difference from the
routine. There wasn't going to be an institution in the world that was going to put
that type of resources into a new type of where genomics was going to be. And so
this was great. The fourth aspect of it is, how can you not like living in Singapore?
Come on, guys. It was just a beautiful city that was going places, and it's Asian.
My wife is not Asian but she is very Asian-oriented. I had kids who were young at
that time. It was an unbelievable opportunity for them to grow up in Asia. They
would never have that opportunity in the US. Asian values, Asian exposure and
so forth. So on a personal level, this was a great adventure for the family, a great
opportunity for the family to reconnect with their Asian roots, for me to reconnect, and for me to do science but at the same time do it in a way that I knew if
I were successful then, because Singapore is a sovereign nation, no matter how

(*Continued*)

scientists in Singapore institutes did was to create a credible research base in the
international research community from the word "Go". Coupled with complementary institutes that brought translation and clinical research to the basic
research, BMRC institutes seeded the transformation of the biologics industry
in Singapore.

(Continued)

small, a success in a sovereign nation will be copied by other sovereign nations. So it was a way to leverage and experiment so that my idea where genomics should be could be heard throughout the world. And that's exactly what happened. If I were just another research institute in Canada or the UK or the US, I would be just another research institute because there's MIT, there's Stanford, there's a bunch of places. But to be successful in Singapore which is the nation of Singapore and successful in a way that not only did good science but actually helped in an economic sector would be something. I knew that all aspirant countries, Korea, China, would actually be looking into this as a model. I knew that. And sure enough, that's exactly what happened. So for all those reasons, personal reasons, professional reasons, intellectual reasons, social justice, social commitment, I really enjoyed the fact that we could as scientists make a difference to a nation.

I did that in Ireland. That was my first taste of how being a scientist could help the peace process and the reason is that I started a National Cancer Institute-Ireland consortium in cancer research that still exists today. It was just at that time that the Good Friday Accord was completed, and they were looking around for a project that both sides could agree on and won't fight about, and that was cancer research. And we helped build Northern Ireland from a backwaters to one of the most powerful places in cancer research. And it was extremely exciting and very gratifying that science could actually make a difference in the political process and make a difference in the lives of cancer patients simultaneously. So I was hoping we could do that in Asia. And I think we succeeded.

We became known as thought leaders within Asia. I was the president of the Human Genome Organisation. We did the Pan-Asian SNP initiative as a leader. Singapore was the leader to bring everybody together, all the Asian countries including China to work together to study the migration of Asians through genomics.[10]

---

[10] Pan-SNPdb: The Pan-Asian SNP Genotyping Database. http://journals.plos.org/plosone/article?id=10.1371/journal.pone.0021451. Retrieved 16 May 2015.

(Continued)

Joint Council Office: The Multidisciplinary Road

Another challenge that BMRC faced was to get people to break out of their silos to work together in bigger platforms, platforms that are internationally competitive. Seen from certain perspectives, Singapore has the disadvantage of being

(Continued)

There was scientific leadership. We were publishing in *Science, Nature, Cell*. We were recognised on a global scale and in terms of that leadership. The third phase is what I would call the transition phase. I already knew that it was time for me to groom the next generation and we started to move many of our processes into translational sciences as soon as possible. We're working with the clinicians, with some more diagnostics, working with industry. I really wanted to have a Singaporean succeed me and so in that process, we identified Ng Huck Hui as a really great successor. And we worked towards that transition. So there was a launch phase, there was thought leadership growth phase, there was actually a stable phase to convert to the next generation of GIS and so it really worked out very well. I think for me, it certainly did.

For me the highlights were when we started publishing in all the key journals, when I won the President's Award for the SARS thing, when GIS transitioned to Ng Huck Hui. They were profound to me on the basis of the fact that each one of them represented milestones of high impact of this organisation to the nation, to the region, recognised to the world, and a transition to non-foreign leadership. I knew for a fact that my role was a transition role. They hired a foreign talent. I became PR here and I hope that I became quite integrated with the society here but that's different from somebody who served in national service (NS). You know this country has reached maturity when the next generation of leaders are also recognised as successful scientists. And that's what happened and that was really a highlight of my time here.

very small. To amount to anything, it has to compete globally but to do that the platforms of research have to have size in order to have clout. One strategy to break down barriers and get people to work together was first to get them to start talking to each other. A*STAR added an office in addition to SERC and BMRC and this was the Joint Council Office (JCO), realising one of A*STAR's earliest goals of promoting Integrative Sciences, bridging the research in SERC and BMRC. Mooted by A*STAR Chairman Lim Chuan Poh who became Chairman in 2007, the Joint Council Office was formed in 2007 to facilitate collaborations between the institutes, including running grant calls for multidisciplinary research projects and organising an annual scientific conference to bring the research community together. Multidisciplinary research that leverages on the entire spectrum of research capabilities in A*STAR has been one of its aims but the challenge was how to make it work well. Conferences being occasions when people from different places get together to exchange views and experiences and to connect, the first annual scientific conference for the research community in Singapore was organised in 2008. In the A*STAR commemorative publication, Lim said: "This provided a rare opportunity for scientists from the biomedical community to meet the scientists and engineers from the physical science and engineering community. Senior scientists told me 'I didn't know A*STAR had so many different capabilities.' These initiatives were designed to encourage cross-council multidisciplinary collaborations. For instance, in the field of medical diagnostic, collaboration between the Genome Institute of Singapore and the Institute of Microelectronics led to the development of very powerful diagnostic devices." The conferences and workshops developed a lot of goodwill in the research community in Singapore. Today, this internal conference is an annual event in the A*STAR calender.

Getting researchers in different parts of the research landscape to talk to each other was to address the issue of researchers working in silos and to build a more collaborative attitude in the community. Networking brought together university faculty members, A*STAR researchers and hospitals to work on projects. Grant calls further shaped collaborative behaviour. BMRC grant calls are open to all researchers in the country. To promote collaboration, A*STAR added a feature called the "Glue Fund" whereby more money will be given to those who work with people outside their domain. For example, if a hospital researcher teams up with someone in the university more funding will kick in. Good collaborative research cannot be forced but it can be encouraged with extra funding support, extra manpower, extra access to equipment and other resources. Lim elaborates: "The Joint Council is a very impactful instrument in another way. If there is a new area of research we want to go into that needs to draw on research capabilities from both the research councils, such as in the biomass conversion space or Medical Technology, the Joint Council can invite proposals in that space from the entire A*STAR research community. The researchers will then self-organise to put forward their proposals. We have the flexibility to convene a fit-for-purpose panel to review the proposals with an integrative and holistic perspective. If it's competitive, relevant and impactful, we'll support it."

The ability to bring together different capabilities from the two councils increases A*STAR's value proposition to industry. This capability connectivity is further fostered by Singapore's small size which simplifies physical connectivity. Fusionopolis that houses the SERC institutes and Biopolis that houses the BMRC institutes are literally right next to each other. The different universities in Singapore are between 10 and 45 minutes apart from Biopolis and Fusionopolis by the mass rapid transit. It underlines the ease of connectivity in Singapore.

The Biomedical Science Journey
Dr Benjamin Seet, Executive Director, BMRC

You see four phases of the Biopolis: creation, then translation into the clinic. The third phase is working with industry, and then the fourth one is creating successful Singaporean enterprises. And these 20 years from 2000 to 2020 will define Singapore's biomedical science journey.

The first phase was the formative phase, where Philip Yeo, Dr Christopher Tan, Dr Kong Hwai Loong, and the pioneer directors and researchers played very big roles to build up science in a place where that level of science never existed before. It was about planting seeds. It was about attracting prominent scientists from around the world, whales as Philip Yeo calls them. IMCB has been around since 1985 but beyond IMCB all the hospitals were primarily providing clinical services and the universities were providing undergraduate education. The only basic research was conducted in IMCB. So the first phase, 2000 to 2005, was about growth and creation, and the focus was to attract good people. In the second phase, five more institutes were created and that gave a certain critical mass of research in Singapore. But they realised that just doing good science and publishing papers was not enough to grow the industry and

(Continued)

Integrative sciences is the trend today because splitting nature into two domains of science is not realistic. A great example of integrative sciences is the Institute of Bioengineering and Nanotechnology, the world's first such research institute. The research output here is often a physical item, a chip, a kit, a physical engineered item, often with a starting point in the biosciences. The researchers use the raw material and model system of the life sciences refined through engineering processes that are quite well-defined and predictable to produce a reliable product for humans. Bioengineering has clearly shown how biology and engineering can work together.

Change in Funding Framework

By 2010, the BMS initiative had been progressing for 10 years. Questions were beginning to be asked about the health and economic benefits from the billions of dollars invested. Industry collaboration plays an important role in not only enlarging the research platform but also keeping research grounded and in line with A*STAR's mission. In 2011, there was a tweaking of the funding framework to get more industry collaboration. The tweak involved reducing

(Continued)

economy. There was a need for translational research in the clinics and hospitals. So from 2005 to 2010 it was very much focused on building up the medical research capabilities. We went from one medical school to today's three medical schools. Today we have academic medical centres where in the past they were education centres. So it's a lot more research intensive whether you go to Outram, Kent Ridge, and in future, Novena Tan Tock Seng. There are new multi-storey blocks dedicated to research and that bring research into the clinical environment so there's more translation to the patient. The other focus will be on training as well as attracting clinician-scientists. That's a breed of scientists that are half doctor, half scientist, people with an MBBS as well as a PhD who work in the healthcare environment and combine it with research. So that was the second phase, 2005 to 2010. We grew from one to 10 institutes and built great translational capabilities in the hospitals. This brings us to the third phase, where the BMRC institutes began to work more closely with industry. It needed new leadership. It needed cultural change and a different way of funding research. It needed diversification of BMRC's research portfolio to work with new industry clusters. But in the end, we exceeded all targets and surpassed every expectation. (For the third phase, see Chapter 7; for the fourth phase, see Chapter 9)

the core budget of the institutes and increasing the proportion of private sector funding through collaborations. Between 1985 and 2010, NSTB/A*STAR had treated biomedical research as investment into basic and long-term research capability. The 2011 change in funding framework required institutes and researchers to raise the value of industry-backed research in BMS R&D. Industry contributions to BMRC research institutes amounted to a mere $18 million from 2006 to 2010. However, between 2011 and 2015, industry spending with BMRC institutes increased many-fold, to eventually exceed $360 million. Such industry-linked projects were to count as a Key Performance Indicator (KPI) of the institutes. Not only was this target of $360 million a quantum leap, it also required a change of cultural mindset and 2011 put BMRC on a new learning curve.

However, the change was not proposed in a vacuum. In the few years preceding 2011 there had been a noticeable shift in the way the pharmaceutical industry was approaching R&D. With the lack of new drugs coming to market and impending expiry of patents, the big pharma companies were downsizing their R&D teams and moving towards a collaborative R&D approach with academia. There was also a shift to the Asia-Pacific region and the time was ripe for A*STAR research institutes to respond to this shift. During the transition, A*STAR Chairman Lim Chuan Poh and BMRC executive director (ED) Prof Lee Eng Hin met with the EDs of the biomedical research institutes on numerous occasions to clarify the intentions and to allay fears. The departures of a couple of high-profile researchers after the announcement of the tweak had prompted much unnecessary international speculation and discussion on where Singapore biomedical research was headed. Prof Swain who founded SICS pointed out in an oral history interview in 2015 that there were institutes such as SICS, a translational institute, that had no issues with the tweak in the funding framework as such institutes routinely collaborate with industry. The message that was clearly articulated was that A*STAR research institutes require a spectrum of capabilities with good fundamental science that will produce the discoveries that can then be translated and eventually commercialised.

The storm in a teacup has since blown over and BMRC has emerged stronger and more industry-ready than ever. There is strong industry collaboration, there is strong collaborative research with hospitals and universities, and the quality of science has strengthened. Said executive director of BMRC Dr Benjamin Seet who came into A*STAR in 2011: "Dealing with the change in mindsets when it comes to working with industry was probably the first challenge, but we realised that at the end of the day, all that really mattered was good science. The science has to be as best as we can make it. It has to be published in the best journals. It has to be reportable in the scientific community to be of any interest to companies. Big companies are not interested in substandard science. They want the best science, the best scientists, the best knowledge leaders. We need all of this to attract companies whatever the industry we work with, particularly the multinationals. Big companies can shop around the world for scientists and for ideas. So we have to be as good as the best centres. So in a sense, while the concern that we would be diluting our science if we worked with industry was very real in 2010, it has not happened. In fact, it is our best scientists who publish in the best journals who are the ones that are of greatest interest to companies."

Says EDB's Managing Director Yeoh Keat Chuan[11]: "Today Singapore is recognised as a leading manufacturing location for pharmaceuticals and biologics as well as various medical technology products ranging from life science tools to medical devices. There is also a growing base of research activities

---

[11] The Biopolis Story: Commemorating 10 Years of Excellence. Singapore, A*STAR and JTC Corporation, 2013.

as companies look to expand their range of products for Asia. Singapore is also a leading location for commercial operations, with seven of the top 10 pharmaceutical companies and all top 10 medical technology companies having regional or global commercial operations based in Singapore." The attraction is Singapore's strong IP regime, its pro-business environment, and its strengthening research capability and talent pool.

Research Integrity
Prof Lee Eng Hin

Today, research integrity has become extremely important. With so many researchers wanting to get ahead there have been news reports of researchers who manipulate their research data in various ways. It is not unknown for a researcher to attempt to re-publish the same data in another journal. Or to make claims that cannot be replicated or validated by other research labs. One way to combat research fraud is through vigilance. We now have research integrity offices in all our universities and in A*STAR too. Anybody who does anything that is ethically unacceptable may be brought up before the research integrity office whose job is to check the integrity of that research and the researcher.

Another way research fraud can occur is when a big name researcher is persuaded to lend his or her name to a publication that he or she may not have actually worked on. This can happen particularly with publications arising from research sponsored by industry in which a big name in research is invited to be a "collaborator" to give the publication more credibility. Obviously this is unethical. What makes research fraud a very limited enterprise today is the number of eyes constantly examining the research data. There are people out there who look at publications very thoroughly and pick out the ones that may have errors, deliberate or otherwise, and they will then bring this up to the researcher's institution.

Maintaining a reputation for both research integrity and research excellence is particularly important for Singapore's research capability to continue to be a value-add proposition for companies wanting to collaborate with us in research and development. Such a reputation for integrity is just as important when we talk about clinical trials because the safety of the patients who participate is paramount. We have hospital ethics boards (Institutional Review Boards) to examine the parameters of such trials to ensure that the patients' safety is not compromised in anyway. Such ethical concerns extend to the use of animals in research. This is governed by an animal ethics committee.

- *Prof Lee has been involved with the research integrity offices of both NUS and A*STAR since they were established.*

Chapter 6

Developing Research-Intensive Universities

Barry Halliwell and Bertil Andersson

Whatever critics may think of global rankings, it is of some significance that Singapore's two major universities, National University of Singapore (NUS) and Nanyang Technological University (NTU) both feature in the Quacquarelli Symonds (QS) World Rankings in the top 40 universities in the world. Such rankings are based on research publication in benchmark peer-reviewed scientific journals, reputation among global academics, etc. It is a significant achievement for a small country that only started on the journey to research excellence with the establishment of the National Science and Technology Board (NSTB) in 1991, the same year as NTU's establishment. Although other measures of excellence must also be taken into account, what they do indicate is the effectiveness of the national investment in education and research as part of the Singapore Government's overall commitment to create a vibrant knowledge-based economy. Without such a commitment, building up research-intensive universities would not have been possible. In tandem with the mission-oriented RIs under NSTB/A*STAR, the universities play important roles in the R&D eco-system. They came in for more attention as A*STAR and the National Research Foundation (NRF) started to spearhead thematic research programmes and fund the universities on emerging science and engineering topics and in areas with potential to become upstream resources for the RIs further downstream. Already an important part of the R&D landscape since the earliest days of any kind of interest in science and technology, the universities are today an even more important component of the R&D hub plan with the creation of the Research, Innovation and Enterprise Council (RIEC) and the NRF. When NRF started to look at emerging areas and embarking on this new strategy of funding new and strategic basic research, the universities came into focus. Unlike A*STAR, NRF is essentially a funding agency and does not manage any research institutes. NRF funds R&D projects that are expected to have longer-term national impact. Thus in 2006 one of the first NRF programmes was $500 million for water research, a sustainable and secure water supply being a continuing national issue, notwithstanding the success of NEWater, a PUB-initiated project in the late 1990s to look into recycling water. Another project in which NRF has invested funding is research into how dengue fever spreads. The National Environment Agency manages the project and turns to the universities and research institutes to learn more about ways to prevent and mitigate the spread of dengue, the diagnosis and treatment options, identification of virus types and distinguishing symptoms. The universities are in the thick of such research projects.

Since 2006 and the formation of NRF, substantial funding for new areas is being set aside to boost basic research in the research-intensive universities in Singapore. In 2005, the way the universities were organised and funded changed. NUS and NTU became autonomous universities with the goal of enabling them to better respond to the opportunities and challenges, as well as to pursue excellence. Until then, NUS and NTU were government statutory boards and subjected to government regulations and practices. The universities became autonomous institutions with their own pro-active Boards of Trustees each directly responsible for the policy for its own institution. The financing of the institutions was also tweaked to give each university more autonomy to decide what it should focus on and the direction of its development. Today, increased administrative and financial autonomy give the universities the space to differentiate themselves and chart their own strategies towards achieving peaks of excellence. Such autonomy enables them to explore different ways to build up teaching and research excellence, raise their international standing and enhance students' experience.

Speaking on the deregulation, former Permanent Secretary of the Ministry of Education and current A*STAR Chairman Lim Chuan Poh said in his keynote address at a Times Higher Education World Summit at NTU in 2013: "The university autonomy review was immediately followed by the review to transform our autonomous universities into globally competitive research-intensive universities. The main recommendations of the review were the significant increase in the Academic Research Fund, the setting up of an international Academic Research Council, and the launching of Research Centres of Excellence or RCEs by the National Research Foundation (NRF) together with the Ministry of Education (MOE). The integration of the RCEs with the universities is aimed at ensuring a cross-flow of talent. RCEs would also increase the capacity of the universities to train more high-quality post-graduate students. Following the review, MOE's research funding for the autonomous universities nearly tripled. On top of MOE stepped-up funding, the universities could compete for R&D funding from A*STAR, NRF and several other public agencies. As a result, the R&D spending at higher education institutes in Singapore almost doubled from $580 million annually in 2006, before the research landscape review, to over $1 billion in 2011.

"This focused approach to research funding not only serves to raise the research intensity within the university, more importantly, it also creates deep capabilities and excellence in selected fields to be among the best in the world. They are certainly in a better position now to work with partners overseas as well as with industry and government to tackle some of the greatest challenges our world faces today and to help shape a brighter future for all of us."

The need to boost basic research in the universities was also articulated in NSTB's second five-year National Science & Technology Plan (1996–2000). In this second plan, NSTB established the Temasek Professorship (TP) Programme as a platform where renowned international R&D research leaders could be invited to lead strategic research projects identified and funded by NSTB as critical to Singapore's scientific and economic development. The TP Programme served as a magnet to draw excellent local and foreign researchers and students to engage in research activities in both NUS and NTU.

The first TP appointee, in 1999, was Prof David Ewins from Imperial College London. He was hosted by NTU to develop a programme on the mechanics of microsystems. The other Temasek Professors who were appointed during the life of the programme were Prof Daniel Wang from MIT, Prof Wolfgang Knoll from the Max Planck Society, Prof Dim-Lee Kwong from the University of Texas, Prof Charanjit Singh Bhatia from IBM Almaden, Prof Ellis Johnson and Prof Rao Tummala from Georgia Institute of Technology, Prof Artur Ekert from Oxford, and Prof Anthony Fane from the University of New South Wales. Besides Prof Ewins, Prof Fane was also hosted at NTU. The remaining TPs were hosted at NUS.

The TP programme achieved a significant level of prominence, visibility and recognition in the international scientific community. This is evidenced by the level of interest it attracted and the stature of the TPs who were appointed. A distinctive feature of the programme was that each TP was expected to commit at least three months per year to a physical presence in Singapore. It is worth noting that in all cases, initial reservations about this time commitment melted away as the opportunity to create a new research team in Singapore, that complemented each TP's research team at home, came to be seen as a workable and significant advantage of the funding.

As a talent attraction programme, the TP scheme was enormously successful. Prof Ewins continues to be associated with NTU. Prof Daniel Wang was actively involved in the development of the Bioprocessing Technology Institute in A*STAR and chaired its Scientific Advisory Board till recently. Following the completion of his TP term, Prof Knoll led a follow-on programme at the Institute for Materials Research and Engineering (IMRE) in A*STAR and remains a Visiting Scientist at NTU. Prof Bhatia Singh took up a tenured Professorship at NUS. Prof Kwong first became the Chair of the Scientific Advisory Board and subsequently executive director at the A*STAR Institute of Microelectronics. Prof Fane set up a large effort in membrane technology at NTU and became founding director of the Singapore Membrane Technology Centre in 2008, funded by the NRF Environment and Water Industry Programme. He remains Director-mentor of the Centre. Prof Ekert grew his initial TP in quantum information technology at NUS, funded by A*STAR, into a winning bid for the first of five Research Centres of Excellence (RCE) funded by NRF and MOE; he has become the founding director of this RCE, the Centre for Quantum Technologies (CQT), since 2007. This early strategy to increase research intensity in the universities through the TP Programme has been followed by the setting up of the Research Centres of Excellence in the universities.

Research Centres of Excellence

Started by the NRF together with MOE, the Research Centres of Excellence (RCEs) are attached to the universities and designed to build up strong research programmes that capitalise on the universities' original research strengths by an infusion of substantial funding. Headed by a distinguished scientist as Director who is advised by an international Scientific Advisory Board, each RCE has significant autonomy in pursuing its own research mission and objectives with a governing board to give strategic direction and stewardship. Each RCE has 15 to 25 principal investigators (PIs), each of whom leads a research team of post-doctoral fellows, research students and supporting staff. All RCE PIs hold joint-faculty appointments either at the host university or in one of the other local universities. The five RCEs were selected through a competitive process between 2007 and 2010. White Papers and subsequently full proposals were submitted for international peer reviews and evaluated by the MOE's Academic Research Council, a committee of distinguished international scientists and academics. The RCE's progress is reviewed by an International Review Panel every three years.

The first RCE in Singapore is the Centre for Quantum Technologies (CQT) at NUS which was started in December 2007. CQT's experimental labs use lasers to manipulate, measure and contain quantum systems containing atoms, as well as conduct experiments exploring the quantum behaviour of light.

Headed by Prof Artur Ekert, who also spends some time as Professor of Quantum Physics at the Mathematical Institute, University of Oxford, CQT brings together quantum physicists and computer scientists to explore the quantum nature of reality and new possibilities in information processing. Research at CQT is focused on understanding and controlling the interactions of atoms and photons. These interactions can potentially be exploited to access new modes of computation and communication. The Centre on the Kent Ridge campus of NUS has more than 200 staff and students from more than 35 countries.

The second RCE is the Cancer Science Institute of Singapore (CSI Singapore) that was officially started in October 2008 and is based in NUS. The Centre aims to catalyse world-class research in cancer sciences by building on prevailing academic strengths and research competencies, attracting top global academic cancer research talent, fostering outstanding international scientific collaborations, and offering quality training to nurture innovative cancer researchers. The Institute takes a multi-faceted and coordinated approach to cancer research extending from basic mechanistic studies all the way to experimental therapeutics. It links unique resources in Singapore to develop new approaches to understand and treat this complex disease. The Institute is headed by Prof Daniel Tenen from Beth Israel Deaconess Medical Centre, Harvard Medical School.

The third RCE is the Earth Observatory of Singapore (EOS) which is based in NTU and started up in December 2008. Through fundamental research on earthquakes, volcanic eruptions, tsunamis, changing sea levels, and climate change in Southeast Asia, EOS seeks to inspire and enable safer and more sustainable societies in the region. Prof Kerry Sieh, who was previously the Robert P Sharp Professor of Geology at the Tectonic Observatory, California Institute of Technology, leads the EOS team of renowned earth scientists.

The fourth RCE is the Mechanobiology Institute (MBI Singapore) which became operational in April 2009 and is hosted by NUS. Its aims are to develop a new paradigm for studying living systems and diseases by focusing on cell and tissue mechanics and to create a common international standard for defining new computational models, experimental reagents and tools for studying diseases of cells and tissues. This Institute is led by Prof Michael Sheetz who was William R Kenan Jr Professor of Columbia University.

The fifth RCE is the Singapore Centre on Environmental Life Sciences Engineering (SCELSE) hosted by NTU in partnership with NUS. It became operational in 2011. It aims to deliver practical solutions to environmental challenges based on complex microbial communities organised as biofilms and using

Fig. 6.1.   Launch of Centre on Environmental Life Sciences Engineering.

state of the art metagenomics. The Director of the Centre is Prof Staffan Kjelleberg from University of New South Wales, Australia, and its Deputy Director is Prof Yehuda Cohen from Hebrew University of Jerusalem, Israel. (See Figure 6.1)

The idea behind the RCEs is that by investing in a research area where the university is good, it can go from good to excellent. The aim is to build a cluster of world-class research centres with PIs publishing their work in the leading scientific journals such as *Nature, Science, Nature Medicine, Nature Biofilms and Microbiomes*. These Centres are to be closely reviewed at the end of the first five years to ensure that the research is of the highest international standard and globally recognised. To start off, the Centres were given a block of, on average, $150 million over 10 years to make themselves excellent. But as the Centres get to Year Five and Six, they have to go out and compete for available research money. What will happen is that after Year 10, while they will continue to get a core grant from NRF, they will have to raise substantial research money from other external sources such as foundations, foreign governments, and external competitive funding schemes. At the end of the 10-year funding, the goal is to have research centres whose quality of research is of such a high standard that they are able to attract such research grants. At the same time, these Centres will be integrated with the universities that have been incubating them. Although the research staff of the Centres are considered mainly as researchers, there is a small teaching load consisting mainly of mentoring and PhD education. All the RCEs have strong international review panels that assess them every three to five years. If a Centre gets a poor report, the primary funding agencies — NRF and MOE — have the right to terminate funding which effectively means shutting down the Centre. However, all the indications are that the RCEs are doing good work and in all likelihood will remain in the Singapore R&D landscape.

Developing a Vibrant Research Culture

The RCEs have been made possible by the ability of the universities to attract and keep good people. This is a role that the universities have played since the beginning of the national R&D push in 1991. They are even more effective in this role because of the existence of these and similar centres but also because global rankings for the universities are higher than what they once were. Prof Hang, former Deputy Chairman of NSTB and A*STAR, who has considerable experience in the task of recruiting research talent says: "I would say Singapore's research changed from a Third World to a First World standard from 1991 to 2005. Usually other countries would take 20 to 30 years to be able to build that kind of a track record. And we did it in 10 to 15 years. One of the major reasons we were able to do that was our foreign talent attraction. We actually used NUS and NTU as the basis of talent attraction. Because if we were to just build up independent research institutes, nobody knew whether the institutes would still be up and running five years down the road because all budgets were usually for five years. Whereas if they were parked in the universities, university people knew that the institutes would continue to exist for a much longer time if they did good work. And even if the institute gets closed down, the key people would get absorbed into the pool of academic staff. So they felt safer to relocate and that made the job of recruiting much easier. Because when we had to recruit the senior people to become executive directors of our research institutes or be senior staff, using the universities to recruit them made it easier. And of course the local academic staff would also participate so that this was win-win for both the institute and the universities." Singapore being English-speaking and at the crossroads between East and West, and as well as having an excellent reputation built on the NSTB/A*STAR research institutes and the universities, has today developed a talent pool of PhD manpower made up of many nationalities. So the universities have become very vibrant institutions. People are taking notice. Visiting professors come here willingly and leave with very good impressions, and frequently make repeat visits.

Another way to increase research intensity in the universities has been the introduction of CREATE, a unique system to import world-leading institutions to Singapore and seed research and enterprise. (See Chapter 3) All the local research-intensive universities work closely with these institutions on projects substantially funded by NRF.

There are three key elements in developing a vibrant research culture. One is people and so the biggest hindrance is not getting the right people. Singapore's drop in the birth rate is going to translate to fewer people, and the increasing anti-immigrant mood seen not just in Singapore but also globally, is a worry. Balancing these political concerns while maintaining a good flow of scientific and engineering talent to the universities and research institutes is another key issue. The second issue is if the investment in R&D shrinks or dries up but this is not yet an issue in Singapore nor is it likely to be. The third issue is if there is too much pressure to do only applied or commercialised research. While a certain degree of applied or commercialised research is good and highlights the universities' links with the economy and society in general, pushing universities too far in that direction may lead them into very short-term activities such as contract research. University research must also move into new areas to meet yet unknown or barely acknowledged challenges and not just focus on current S&T problems. History has shown that some of the advances in today's medical science have come from outside medical research. For example, laser technology, now widely used in surgery, came originally from research in physics. The idea of stem cells originated in botanical research. While the use of stem cells for medical treatment is still in its infancy, it is one that shows much promise. Concentrating only on applied research may lead to missed opportunities and the history of science and technology is replete with such anecdotes. While university research should never be too use-inspired or too applied, at the same time a small country like Singapore cannot afford to have 3,000 academics all doing blue sky research (defined as research into areas where applications are not immediately apparent) because there is a national mission to consider, quite apart from funding being mostly from tax payers' money and with political accountability to think about. It is a balancing act for university research administrators and advisory panels to decide where to invest research money and how much to invest in basic research.

Choosing Research Priorities

Where research areas are initially decided by individual inspiration and passion, today universities and research organisations decide on funding for research through a competitive grant mechanism, while maintaining individual research freedom from which new areas will inevitably emerge. Researchers respond to grant calls with research project proposals. Such proposals may be submitted to different agencies and universities and be taken up or rejected as the case may be. The big advantage of a competitive grant mechanism is that it allows R&D managers to shape the research agenda very quickly. So if a country wants more research on water, more money is given to such research projects. If it stops putting money into such research, then the research will shrink back to just the core work of passionate academics. However, the competitive grant mechanism must be used with care and discrimination. It can lead to an over-emphasis on research with more immediate economic gains and thus future trends may be missed. It can build up a body of expertise in a certain area that is then lost because funding has dried up when the competitive grant mechanism moves research interest into a different area.

As in the past, what is researched is often personal inspiration and passion whether it be blue sky research or even new applications for science whose economic value is still unknown. Such early work may then lead to applications for research grants and the expansion of a new research area. Such personal inspiration and passion is part of the story of Singapore research and its research institutes, and it must continue to play an important role in uncovering new research areas whose economic potential is still unknown. Without such personal inspiration the story of NEWater and its role in Singapore's water sustainability would be a different one. NEWater's origins rely on research that was first done in PUB, then NUS and NTU all in the time before water sustainability became a national issue. This is another story about research and research impact. It is a two-fold process. First, there are always a few academics working on esoteric topics when somewhere a decision is made that more research is needed in that area because a need has been identified. More money is then invested in such research, thus drawing more researchers into the area. Academics are always looking for research money because it is a competitive field. So the one or two scientists working in that field are enabled with more research funding to get a bigger lab, train a lot more PhD students, hire more post-doctoral fellows and develop something impactful from what appeared esoteric in the beginning.

In the case of Singapore, the identification of need often comes from the EDB. Charged with Singapore's economic development including attracting foreign inward investment, EDB works closely with the universities and A*STAR to develop research. What the areas of academic research sponsored by EDB should be are determined in large part by national needs.

The universities do not go into a particular area of research because it has the capability but rather they develop the capability because the country needs it. Says Prof Hang from his extensive experience in NSTB and A*STAR as well as the first Director of Research in NUS: "We first identified the need, then we went to the university and asked 'Do we have the capability? If we do not have, then we import it. If we have the local expertise, we will use it and try to grow it.' So that was the way EDB, NSTB and now A*STAR works. EDB identifies the needs by talking to multinational corporation (MNC) partners and says 'What would it take for you to come to Singapore? Do you need R&D? If you need R&D, yes, we would like to help you to set these up. We will have grants and tax breaks to help you to lower your start-up costs.' It's a very pragmatic approach but in the process, the university professors and PhD students get access to really state-of-the-art equipment, new ideas, new processes. They become involved in developing these new ideas and processes. This was how it began. But in the long run, as industries were attracted to set up R&D in Singapore, more and more of the more R&D would move to the companies. Then the universities would find resources to do more basic research, longer-term research in order to support the more mid and downstream R&D. In Singapore's case, we were very pragmatic. In general, we developed the more mission-oriented areas first and then five years, 10 years, 15 years down the line, we move more towards longer-term basic research. But it is still economically relevant. It is still basic research to support what we have built up over the years. The very, very new areas of research that are being done in the universities today was built on the pragmatism of the earlier years."

As a more structured way to encourage companies to partner universities with co-funding from NRF to do longer-term research in Singapore which is relevant to their future needs, the NRF established the Corporate Laboratory@ University Scheme in 2013. It seeks to attract foreign and local companies to collaborate with local universities on industry-relevant research that will enable researchers, academic staff, PhD and post-doctoral students to work alongside companies on programmes that have direct relevance in the marketplace. For instance, the Rolls-Royce@NTU Corporate Laboratory will focus on three areas of experimental research: Electrical Power and Control Systems, Manufacturing and Repair Technologies and Computational Engineering. The Keppel-NUS Corporate Laboratory will focus on three research thrusts: Future Systems (Deepwater Technology and Arctic Technology), Future Yards (Productivity Enhancement of Yard Operations) and Future Resources (Deepsea Seabed Nodule Collection). (See Figures 6.2 and 6.3)

Fig. 6.2.   Launch of Keppel-NUS Corporate Research Lab.

Fig. 6.3.   Launch of Rolls-Royce@NTU Corporate Research Lab.

National University of Singapore

The seeds for developing a stronger research culture in Singapore's universities had been sown in 1980 with the creation of the National University of Singapore (NUS) from a merger of University of Singapore (founded in 1949 as University of Malaya in Singapore) with Nanyang University (founded in 1956 as a Chinese-medium university), and the setting up of Nanyang Technological Institute (NTI) within NUS but based on the Jurong campus of Nanyang University. Dr Tony Tan, the first Vice-Chancellor (VC) of NUS (the position is now titled President) was VC for a year before becoming Minister for Trade and Industry from June 1981. He spent his year as VC identifying the priorities and directions of the university. According to *Beyond Degrees*, a book on the history of NUS: "As he saw it, the first and foremost role of the NUS was to produce graduates equipped to satisfy the manpower needs of the country. Singapore was in the process of ushering in rapid economic change, and it was felt that the most urgent task was to produce the managerial, professional, technological and scientific manpower to further Singapore's social and economic development. In defining this basic function of a national university, the Vice-Chancellor was reiterating the basic commitment of the University to that of meeting the needs of an expanding community and developing nation." As reported in the same NUS history, speaking at the Students' Union Welcome Convention on 20 May 1980, Dr Tan said: "Our economy has an insatiable demand for technological and professional manpower. For the present I do not see any escape from the necessity to gear university education to the demands of the market.... (O)ur priorities do not permit any other course. Students want it, society needs it and the university should provide it." The next Vice-Chancellor, Prof Lim Pin, who sat on the board of EDB was well-positioned to know what the demands of the market were. He instituted a mechanism to get feedback from employers on their needs so that the university curriculum in all the faculties could be kept constantly updated to meet those needs. As those needs changed and the economic goals became the creation of a knowledge-based economy, thus began the drive to turn NUS and eventually NTU into research-intensive universities while keeping their traditional role as teaching universities.

Three years after Prof CC Hang became the founding Deputy Chairman of NSTB, in 1994 he was asked by Prof Lim Pin to look into beefing up research in the university. Prof Hang said: "In the '80s, NUS was a teaching-oriented university. We had a low level of research that aimed just to keep ourselves abreast of technological changes, developments in new areas. We had a very small number of PhD students in all faculties. We had a very small handful of prominent researchers, for example, Prof SS Ratnam who was world-famous for his reproductive research, Prof Arif Bongso for stem cell research, Prof Lam Toong Jin in Biology. In Engineering at that time, we could only quote one or two persons like Prof Lee Seng Lip who came to Singapore from the US because of his family. He was a very well-known researcher and he managed to continue with his research and consulting. He invented the fibre-drain for use in land reclamation. So there were only a handful of noted researchers. The majority were bogged down by teaching. Academic staff had a very high teaching load as NUS was expanding its enrolment. In addition, there was an absence of PhD students owing to job scarcity, and an absence of large research funding."

The bulk of this early research was supported by funding from the Ministry of Education (MOE), grants in the tens of thousands of dollars. These grants kept alive research in areas that were not yet on the national agenda such as civil engineering, chemical engineering, and water. Meanwhile some of the larger grants were coming from the Ministry of Defence and they went to the Science and Engineering academics. Starting in the mid-1980s the EDB and Science Council began giving bigger research grants for applied research. The few research-active academic staff were pleased to see research grants coming in the few hundreds of thousands for a few years.

After the formation of NSTB in 1991 and the first National Technology Plan, more emphasis was given to the research activities of academic staff. The universities added a second mission, that of progressively becoming research-intensive, on top of their primary mission of teaching. With this new focus on research, the faculties began designating various staff members to handle such research grants. In 1994, NUS beefed up the status of research by creating an Office of Research with Prof Hang as the founding director. He began by talking to all the deans and heads of departments on the new directions for the university. He says: "After those nine or 10 months, I became the Deputy Vice-Chancellor in charge of research and other areas. Because at that time we didn't have the position of Provost, so the Deputy Vice-Chancellor had to multi-task. My portfolio was research, enterprise, finance and also because there was no Provost, the two Deputy Vice-Chancellors each had to look after half of the faculties. So we had regular meetings with the deans and senior management of the faculties that were under our charge. For the next six, seven years I had the opportunity to drive research because the Office of Research was under me. I started to second senior academics to the Office of Research to help strengthen Engineering and Science research but also to help the other faculties such as Architecture, Building, Social Sciences, Dentistry, and other teaching-oriented departments to embrace research. Research should be for all faculties, all departments, because NUS deeply believes in the scholar-teacher model, that teaching and research go hand in hand, and it's a close nexus and this is the way to build the future university. For the older academics it was a very hard time because they had been there teaching for the last 20, 25 years. They were five to 10 years from retirement. How could they start a new engine? So my job at that time was to make sure that they helped us to talk to the young people, to persuade them that the future is in research and teaching, the scholar-teacher model." The Office of Research subsequently became the Deputy President (Research and Technology) Office to oversee the university's research concerns that include attracting funding, strategic research investments, managing research ethics and research safety. This position was not filled when Prof Hang was seconded full-time to NSTB/A*STAR as its Executive Deputy Chairman from 2001 to 2003 until Prof Barry Halliwell was appointed in 2005.

Prof Halliwell first came to Singapore in 1998 on sabbatical as Visiting Professor at the Biochemistry Department of NUS and became head of the Biochemistry Department from 2001 to 2007. Between 2005 until he stepped down from the position of Deputy President at the end of May 2015, he administered the university's research programmes. Recruitment of new academic staff began focusing equally on research quality as well as teaching ability. The mission was to really raise the quality of academic staff and recruit people who were much more research intensive. There was a drive to reform the promotion and tenure system. At that time academic staff who taught well were given tenure as a matter of course. This was changed to emphasise research so staff had to be good in research as well as teach to get promotion and tenure and there were staff who didn't get promotion and tenure. This process of renewal in NUS was made easier by the fact that at that time the retirement age was 55, which created more vacancies for new recruitment. Today, the retirement age has gone from 62 to 65 and now there is talk of 67.

The 21st century also saw the start of a long-term national investment in PhD manpower. The shortage of PhD manpower made it more imperative for Singapore to turn to the universities to groom R&D leaders from the professors who had already made their names in the world of basic research to get them more interested in "use-inspired" basic research. These professors would be the pioneers in bridging the gap between basic and use-inspired research, including what is now called translational research. Prof Halliwell was also founding executive director of the NUS Graduate School for Integrative Sciences and Engineering, created to improve the quality of PhD training, particularly in multidisciplinary research.

The earliest NSTB/A*STAR research institutes incubated first in NUS and NTU were started with the help of experienced research leaders working in MNCs. This strategy could not work for the universities and it was very difficult to recruit high-level people from overseas because Singapore was seen as a backwater then. For instance, the power houses for biomedical research are the United States and some of the European countries like the UK, Germany, to some extent France, with several noted research centres dotted around. Many of the researchers that NUS approached were wary. There was the fear that their peers would forget their existence. So in the beginning it was quite hard to attract the first group of high-level academics. However, once the university succeeds in attracting some very good people, others will notice. One of the people that NUS succeeded in attracting was a Japanese researcher into cancer, Prof Yoshiaki Ito. He had reached retirement age at University of Tokyo but he was still doing excellent research. The Japanese being strict about enforcing such rules as retirement age, Prof Ito retired, came to Singapore, and attracted several notable researchers here in his footsteps.

It was less difficult to attract young academics to Singapore. Singapore's strong point was the government support of research scholarships and research grants. We were able to recruit good people because we were able to say to them: "Look, in other countries you have to write research grants before you can get your PhD students. Here the Ministry of Education gives you the research scholars even before you have had your first research grant." This was Singapore's long-term national investment in PhD manpower. To strengthen research the university also had to grow its PhD student pool. As more funding for more research projects came in, the pool of PhD students and post-doctoral fellows had to grow bigger to support the more senior level researchers. And preferably, the PhD manpower pool that would be developed had to have a strong local core. However, in the 1990s, the PhD student recruitment ran into a wall of incomprehension among Singaporean students. Apart from academia, there was no demand for PhD graduates in industry. The RIs were only just being established. The then NSTB Chairman Teo Ming Kian recalled in 2014 the queries from puzzled students about why he was promoting PhD education. Prof Hang and a team from the Office of Research began visiting universities in China and India to recruit PhD students. The arrival of foreign PhD students who then became attached to NSTB, then A*STAR research institutes, in turn drew in the local students. The selling points for these foreign students were the state-of-the-art equipment in the institutes as well as the strong government support for research. There would be job possibilities after graduation.

As part of the goal of becoming research-intensive, the structure of NUS had to be reorganised. Faculties and departments were studied and restructured. For example, it was found that large parts of the Bachelor of Science curriculum in certain departments such as Anatomy, Physiology and Biochemistry were run independently with considerable duplication of effort and resources. All those programmes with the same basic core curriculum were restructured into a life sciences curriculum with common core courses and electives so that students could choose what they wanted to branch out towards after completing a common core of courses. Teaching facilities were re-organised so that the teaching labs became common user labs instead of each department having its own teaching labs. This more integrated approach not only maximised the usage of resources but also underscored the commonality of the life sciences and more importantly, created more options for students. Such integration also got people talking more with each other and moving out of silos. In the last few years, there has been a massive rise in intellectual vibrancy at NUS. That means people coming together, discussing a topic and then going off to do their experiments. This is not confined to Science and Technology as the same is true in Humanities and Social Sciences. NUS has a number of very strong research centres there. And it is because people from different parts of the university are coming together and talking to each other.

Today, there is a series of networks and programmes so that most faculties have faculty research centres that draw people together from different departments to work on joint research projects. In addition, there are university-level research centres that bring together people from different faculties. For example, the Life Sciences Institute draws together staff from Anatomy, Physiology, Engineering, Computing, etc. Such a strategy has created a culture of people talking to each other. People from Science, Engineering and Medicine are talking to each other to try and use new materials for example, in their respective fields. Although the Internet has been credited with drawing researchers together and in more constant contact than ever before, there is still nothing like the serendipity of face-to-face contact. The teaching load has also been considerably reduced from the heavy loads of the 1990s, dropping from about four or five to about two or three modules a year. At that time, there were some full professors who were tasked with handling some 20 to 30 final-year projects regularly. Today, most supervise about five final-year projects per year. The reduction releases the professors to devote time to research. Such a reduction in the teaching load was made possible by increasing the size of the academic staff to keep up with the rising student enrolment.

The emerging culture of inter-disciplinary research had enabled NUS to set up in 2007, the Interactive and Digital Media Institute (IDMI) bringing together researchers in engineering, physical science and social sciences under one roof, in response to the national call for developing IDM as one of the three new strategic research thrusts under the National Research Foundation. IDMI aims to nurture a new generation of graduate (PhD/Master) researchers doing cutting edge use-inspired research in Interactive and Digital Media, and to transfer the ground-breaking technology that it produces to industry, so that the knowledge may benefit the various facets of society, from entertainment, education and healthcare, to tourism, homeland security, as well as arts and culture. Comprising eight laboratories in areas of mixed reality, sociable robotics, games, ambient intelligence, multimedia sensing, cognitive and social studies, arts and creativity, IDMI has a critical mass of about 200 professors, researchers and graduate students to contribute to this national direction.

Prof Halliwell's Office advises academics on which agency they should go to for their particular project. One of the aims for a more organised funding strategy was to strengthen research in the universities. NUS which is block-funded for education by the Ministry of Education (MOE) has a whole spectrum of fundamental research in areas like mathematics, physics, chemistry, engineering, cell and molecular biology, marine sciences and so on. MOE runs a grant mechanism which specifically targets fundamental research. So if a researcher has a fundamental piece of research, MOE is the likely place to go to. If a researcher were doing fundamental research that has a specific aim of developing a better this or a better that, then the agency to go to might be the NRF. Putting together the different funding agencies gives researchers a broad spectrum of research funding for a range of research. University research funds can also come across borders. Such cross-border funding raises the credibility of the research when it is funded by reputable agencies such as Wellcome Trust, US National Institute of Health, Bill and Melinda Gates Foundation, and so on.

NUS's philosophy in research is essentially that the university should do high quality research across a broad base because one never knows what is going to be useful next. For example, NUS has research in plant biology and ecology, once an area of economically relevant research that became less supported owing to the national focus on biomedical sciences. NUS's research into plant biology is now turning out to be very important because global warming, erosion of coastlines and other environmental issues are impacting plant behaviour, animal behaviour and human welfare. It is now a very important research area and NUS is the only institution here carrying out that kind of research.

This is another example of a broad-based research suddenly becoming important. The NUS School of Medicine is part of a gastric cancer consortium which started with a seed grant of about $70,000 about 10 years ago but it is also a consortium that involves several other faculties including Engineering. With the initial grant the School of Medicine recruited some very high-level investigators to develop a programme. This early focus on gastric cancer led to NUS getting one of the Research Centres of Excellence for research into cancer. The programme has now attracted industry to come and work with it in Singapore. It is developing new techniques to treat gastric cancer and these new technologies are being marketed and are helping people directly. It is a wonderful example of how small beginnings lead to bigger things. And a lot of what is happening in translational research with drugs, diagnostics and so on is because of investments that NUS made in the life sciences 10 to 15 years ago.

The following is yet another unexpected example. A small group of academic staff in the Department of Civil Engineering has been conducting academic research in offshore engineering since the late 1980s even though it was thought to be of marginal importance to the Singapore economy. They persisted and succeeded in convincing the Keppel Corporation and Lloyd's Register to set up Professorships and founded the Centre for Offshore Research & Engineering (CORE) in 2003. Since then, CORE has secured substantial R&D funding from EDB, Maritime and Port Authority (MPA), A*STAR, Singapore Maritime Institute (SMI), and industry. Additional Professorships are supported by MPA and EDB in Maritime Technology, Subsea Engineering and Petroleum Engineering. Postgraduate degree programmes have been established in the departments of Civil and Environmental Engineering and Mechanical Engineering on Offshore Technology, with specialisations in Subsea Engineering and Petroleum Engineering. The NRF further encouraged and supported the formation of the Keppel-NUS Corporate Laboratory with substantial R&D funding in November 2013. NRF and government agencies also invest in a world-class Ocean Basin that will serve as the national focal point for deepwater technology R&D. It is hosted within the NUS campus.

While no university can cover all broad areas of research, it is possible to have "peaks of excellence". The goal is for the universities — and the Research Centres of Excellence — to strive to be among the top five or 10 in the world in certain peaks. Quantum Technologies is one of the peaks for NUS. In an area like cancer research, it is impossible for a small centre like Cancer Science Institute of Singapore to be in the top five in the world in all areas of cancer research. However, it can be internationally known for certain areas such as gastric cancer, and Singapore has become very well-known for its work on gastric cancer. Gastric cancer is much more common in Asia than it is in the US or Europe where all the research was being done until recently. Diseases and cancers common in Asia but less common elsewhere were not receiving much attention in the more research-sophisticated countries with longer histories in biomedical research. Thus, Singapore biomedical research has been able to establish a quick reputation for certain cancers common to Asians by investing early in such research.

Singapore is an island-state with very limited water supply. Prof Neal Chung's research group at the Chemical and Biomolecular Engineering Department in the Faculty of Engineering started to focus on membrane research mainly for water reuse and seawater desalination about 25 years ago. With the expansion of petrochemical and life science industries in Singapore, this group's membrane R&D was extended to chemical, biofuel, energy, pharmaceutical and biomedical areas. In addition to receiving large grants from Government funding agencies, both local and international companies and research institutes also actively partner this group in collaborative membrane research. Today, Singapore's research in membranes for clean water and clean energy has received worldwide recognition. Various membrane technologies have been developed and patented. In 2013 the Lux report from Lux Research USA, the world leading advisory firm providing strategic advice on research intelligence and emerging technologies, ranked NUS as the world's best on water research in terms of membrane research, water reuse and desalination.

Focusing on the Asian phenotype is yet another strategy that is helping to distinguish NUS biomedical research. For example, it is now known that reaction to drugs can differ depending on race. Where previously drug testing was usually done with Caucasians, Hispanics or Blacks, drug trials on Asians, different types of Asians, reveal significantly different results. Different ethnic groups metabolise drugs differently so some drugs that are not effective on certain ethnic groups may turn out to work better on others, or alternatively be more toxic. Singapore has a window of opportunity in this kind of Asian phenotype research because it has the all-important ethnic diversity. If you can say that your drug is being tested on different ethnic groups and you have the effective dosage that each racial group needs, then the marketing can be much more effective. Coupled with ethnic diversity, Singapore's window of opportunity also comes from its international reputation for credibility and probity.

Areas for research need to be revisited periodically because changes in technology, politics, and society among various factors may render previously rejected proposals viable or full of potential. An idea that came up in the late 1980s and which was finally realised in 2001 was the setting up of the NUS Tropical Marine Science Institute. The institute is based in the Kent Ridge campus but it has research facilities on St John's Island. It was persistence over 10 years and the willing support of two Vice-Chancellors that got the institute set up as part of NUS. The institute has multidisciplinary research laboratories and research programmes that focus on policy, environmental and crisis management, primary production and environmental forecasting. Rising sea levels as a consequence of global warming must be of concern to an island-nation like Singapore.

New Areas: Behavioural Sciences

The thinking among some R&D managers, past and present, is that it is time for Singapore to go into its own research into the behavioural sciences. If there is the Asian phenotype when it comes to the biomedical sciences, it is even clearer that Asian cultures differ greatly from each other as well as with non-Asian cultures, thus affecting human behaviour in significant ways. Such differences impact on how technology is used, what business models succeed better, where are the new economic opportunities in different cultural contexts, the impact of technology innovations on life, behaviour and so on. Some R&D managers are of the opinion that the time is ripe for a Social and Economic Science Research Council. Social science research is less expensive because it needs much less equipment. Big Data from the ubiquity of computers and the phenomenal growth of social media are huge resources for social scientists that are currently not well-tapped. The kind of R&D social scientists do is very different from what the scientists and engineers do but is no less valuable and can have tremendous impact on public policy, economic activities and even lifestyles.

Another reason for investing in social sciences research is that technological innovations particularly the digital revolution — Internet, social media, e-commerce, cyber security, digital media, artificial intelligence, robotics, among the key innovations — have all affected human behaviour profoundly and opened up new fields of research. Such innovations have called for new ways of looking at everyday activities and objects — and formulating new policies that affect politics, social behaviour, safety and security, freedom of expression, among many issues. Change and innovation is happening faster than research in these fields. There is a growing interest in the behavioural sciences nationally and NUS is again leading the way.

NRF Chief Executive Officer Low Teck Seng has highlighted a behavioural science research project to study crowd management. Such a research project has been given extra weight by the New Year's Eve tragedy in Shanghai in 2014/2015. It came on the research radar of behavioural scientists in the universities given Singapore's relatively small public spaces and the number of festive events that bring together large crowds. In early 2015, NUS announced the recruitment of a behavioural scientist, Prof Ho Teck Hua, as the successor of Prof Halliwell. An engineer by training, Prof Ho became interested in marketing and went from that field into the behavioural sciences. Prof Ho leaves a permanent position with the University of California, Berkeley, to return to his home country Singapore. Among the research projects that he will be looking into is a study of data from a local hospital to see how patient waiting time can be reduced and how to lessen the chances of a patient being re-admitted within 48 hours. In another project he will examine data from a taxi company to see how the accident rate can be reduced. Speaking on his appointment with *The Straits Times*, Prof Ho said: "Excellent research creates knowledge and transforms lives." NUS's research and teaching philosophy have always aimed for those two goals.

Nanyang Technological University

Nanyang Technological University is a very young university that only started development as a research-intensive university from 2003. That it should be ranked among the top 40 universities despite its youth is a remarkable achievement. But we should not just rely on rankings as the proof of NTU's remarkable progress. Rankings have their critics and it should be remembered that they were established as a guide for future students rather than as a performance measure of the universities. While not dismissing rankings, they should be considered alongside other indicators including impact measures and other bibliometrics which reflect the vitality and relevance of research and innovation. Here, the story is the same with NTU again showing a very rapid rise in these assessment indicators. It now ranks within the top 50 in the world in the *Nature* Publication Index (NPI) and even higher in chemistry and in the physical sciences. This demonstrates the strength of NTU's overall research especially when considering that NTU remains substantially an engineering-based university where researchers generally publish in other journals than those measured by the NPI. It also shows the rising credibility of NTU's academic record as well as its existing achievements in more application-orientated research.

Despite its youthfulness, NTU has maintained the tradition set by its predecessor institution, the Nanyang Technological Institute (NTI), established in 1981 and at that time affiliated to NUS, in terms of its having a strong application orientation. It provides education for practice professionals, embraces innovation as well as having close links with industry, especially advanced technologically-based MNCs. Today, its portfolio of research collaborations with such world leading companies is impressive and the overall income generated has meant that NTU is in the general top category (within the Thompson-Reuters system) of research universities for its industry and innovation income per capita of faculty.

How has this growth and rapid rise in academic, especially research, esteem been achieved? Of course, without the substantial public investment in the higher education sector, the development of NTU, NUS and the other universities, much of this could not have been achieved. It is not only the level of investment which is important but the sustained government commitment and engagement that has created the environment in which the higher education institutions in Singapore can flourish. NTU only really embarked on its journey to become a leading research intensive university after 2003 and in so doing it has also become a more comprehensive institution. Up until that time, it had concentrated its efforts as a predominantly teaching institution.

Development and Expansion

NTU originated as an institution devoted to engineering education. When NTI was created as part of NUS in 1981, it was located on the site of the former Nanyang University (Nantah), the Chinese-medium university that was merged with NUS 10 years earlier. It had three engineering disciplines — civil and structural, electrical and electronic, and mechanical and production engineering. Three engineering schools were added, and the School of Accountancy from NUS was transferred to NTI in 1987 to evolve into the well-respected Nanyang Business School. NTU was established as an independent degree-awarding institution in 1991 with the merger of NTI and the National Institute of Education (NIE), which is now regarded amongst the top 10 of such pedagogical educational establishments in the world.

In 1992, soon after its establishment, NTU added a new school — the Wee Kim Wee School of Communication and Information (named after one of Singapore's former Presidents) — starting its trend to becoming more comprehensive. Midway through its first decade, NTU added the Institute for Defence and Strategic Studies which later evolved in 2007 into the highly-regarded graduate S Rajaratnam School of International Studies (RSIS).

The original NTI had established an applied sciences activity so that in 1991, a natural evolution was to develop part of this into materials sciences. Applied sciences then became the progenitor of two new engineering schools — Materials Science and Engineering and the School of Computer Sciences, both established in 2000. The initial engineering schools then evolved into their three modern equivalents namely, Civil and Environmental Engineering, Electrical and Electronic Engineering and Mechanical and Aerospace Engineering. The major expansion of the university took place in the early years of this century with, first, the creation in 2002 of the School of Biological Sciences, its first move into the natural sciences. This was followed closely by Humanities and Social Sciences in 2004, and the School of Physical and Mathematical Sciences and the expansion of engineering with the School of Chemical and Biomedical Engineering, both in 2005. A more radical addition to the NTU portfolio allowing the university to stretch across the disciplinary spectrum occurred in 2006 with the creation of the School of Art, Design and Media. One of most significant developments in terms of both teaching and research has been the addition, in 2010, of the Lee Kong Chian School of Medicine (LKC Medicine), which is a joint School with Imperial College London offering leading-edge medical education and developing an advanced biomedical research programme. It is Singapore's third medical school to meet the growing needs for healthcare within the country. It opened its doors to the first high-level student intake in 2013. Finally, the latest addition to the range of research and education that the university now provides is the Asian School of Environment created in 2014. All have contributed to NTU's success, perhaps most noteworthy being the Division of Chemistry within the School of Physical and Mathematical Sciences now ranked in the top 10 in the world in the *Nature* Publication Index.

The First Decade (1991–2000)

The first National Technology Plan published in 1991 coincided with the real birth of the university with the transmutation of NTI into NTU, and the research programme increased in parallel with the increase in faculty numbers although research was still not a major element in NTU's progress in those early years. However, the application-based nature of research for which NTU is known was set with the siting of the Singapore Institute of Manufacturing Technology (SIMTech) on the NTU campus. The post-1991 focus on R&D was an important impetus for NTU to move from its previous NTI focus on teaching into a more 'Humboldtian' construct that brings research and advanced teaching together. The general increased investment in R&D in Singapore also saw a quantum leap in investment in the university sector, in 2006 just at the time when NTU began its steady progress into a research-intensive institution. So these two dates represent key step changes in NTU's research development. The university also started to become more international in outlook with both student exchange programmes and the start of international faculty recruitment.

Into the 21st Century (2001–2010)

NTU's development in the first decade of the 21st century paralleled some really significant changes and expansion in the higher education and research landscape and ecosystem of Singapore. This was the time when the universities became autonomous and were thus given the space to plot their own strategies to achieve excellence. The new NTU Board had a vision to continue NTU's journey into research intensity coupled with its more comprehensive disciplinary portfolio. NTU had started out with a British academic structure which was then converted into an American university structure under a President although it was not until 2007 that a full structure with both a President and a Provost was adopted. A major re-organisation of the faculty structure was also undertaken with the creation of four Colleges each headed by a Dean: Engineering, the largest and still 60% of the university; Science; Humanities, Arts and Social Sciences; and the Nanyang Business School. Together with the heads of the autonomous institutes of National Institute of Education and Rajaratnam School of International Studies, and new Vice-Presidential and Associate Provost positions, NTU today has created a modern and powerful academic management structure. The year 2007 saw the arrival of Prof Bertil Andersson, former Chief Executive of the European Science Foundation and Rector of Linköping University in Sweden and a world-renowned plant biochemist from Sweden with a longstanding association with the Nobel Foundation and former Chair of its Chemistry Committee. Prof Andersson, first as Provost and later President of NTU, was charged specifically with developing research within the university to an internationally competitive level. It is a coincidence that Prof Barry Halliwell at NUS and Prof Andersson at NTU are both internationally renowned researchers with overlapping research interests in biochemistry, especially in the areas of toxicology and cellular stress. Prior to coming to Singapore, Prof Andersson had been a member of the NRF scientific advisory panel. NTU's reform and advance is very much due to the leadership of Prof Andersson and his team of senior administrators.

The middle of the decade, when the major management re-organisation of NTU took place, saw the fourth iteration of what had become the Science and Technology Plan for the 2006–2010 period. Funding was increased substantially over earlier plans and with a major part of this funding settlement devoted to the university sector. It provided substantial new opportunities for the researchers in NTU to compete for new funding. Particularly important were the schemes to support the Research Centres of Excellence (RCEs) and the introduction of the NRF Fellowship. This latter was closely modelled on the European Science Foundation (ESF) European Young Investigator Awards (EURYI) scheme, with which Prof Andersson had been closely associated. This scheme is now a key part of the European Research Council.

A further major change took place at that time with the introduction of a new faculty tenure system. This provided NTU with an opportunity to review all its then tenured faculty using the highest level of international best practice with the aim of becoming a truly research-intensive university. It was a very major shake-up of the institution and around one-fifth of the existing tenured staff failed this review process. The result of the tenure review enabled NTU to embark on a two-pronged recruitment drive. One component was the recruitment of top quality senior researchers including members of the US National Academy of Sciences and the Royal Society of London. These senior people thus nucleated powerful research groups and were also able to win new major competitive funding awards. The second strand was the recruitment of very bright young researchers both through the NRF Fellowship scheme and also through the university's own Nanyang Assistant Professorships Scheme. Both programmes enabled the young professors to develop their careers through generous start-up grants. Together with the release of the potential of faculty who had been confirmed by the Appointments, Promotion and Tenure process, these measures have ensured that NTU, through its faculty, can operate at the forefront of research and translate this knowledge into both innovation and its educational programmes in a true 'Humboldtian' spirit.

Another development was to emphasise inter-disciplinary research by the creation of pan-university research institutes which have established themselves as international focal points having a critical inter-disciplinary mass. The outstanding examples are the Nanyang Environment and Water Research Institute (NEWRI) and the Energy Research Institute at NTU (ERI@N). NTU also established an inter-disciplinary Graduate School to take the concept even further in the training of future researchers and it is expected to become the dominant factor in the PhD research programmes.

Teaching was not neglected in this move towards research intensification. A Blue Ribbon Commission was established which reviewed the undergraduate programmes and made many radical recommendations to create a new and modern educational approach making the best use of modern technology and also incorporating the latest pedagogical approaches. It was very much moving from "Gutenberg to Gates". It also incorporated a new approach in which emphasis was given to the whole learning process, social as well as academic.

A New Decade of Progress

As NTU entered the second decade of the 21st century, the above reforms have delivered on its ambitions while at the same time the university has advanced in other areas of endeavour. Perhaps the most significant development has been the initiation of a new Medical School to meet Singapore's need for more physicians as health demands change and increase and the population ages. Established as the Lee Kong Chian (LKC) School of Medicine as a joint activity between NTU and Imperial College London, the aim of the new School is to train a generation of doctors who will put patients at the centre of exemplary medical care. Graduates of the five-year undergraduate medical degree programme that began in 2013 will have a strong understanding of the scientific basis of medicine, along with inter-disciplinary subjects including business management, humanities and technology, and be schooled using the latest and most advanced medical pedagogy. Named after local philanthropist Lee Kong Chian, the School aims to be a future model for innovative medical education. Its first doctors will graduate in 2018 with a Bachelor of Medicine and Bachelor of Surgery (MBBS), awarded jointly by NTU and Imperial College London, and become doctors who will enhance Singapore's healthcare in the decades to come. With the Lee Kong Chian endowment, history has made a full circle as Lee Kong Chian was one of the people behind the foundation of the predecessor Nanyang University on the same campus. The School, like the Colleges, is headed by a Dean who thus completes the line-up of NTU's top management.

Both Imperial College London and NTU have the ability to combine medical research and engineering as well as developing biomedical research *per se.* Bringing together biomedical research with engineering holds great prospects for significant advances in research and its application in healthcare services. LKC has recruited top international researchers in clinical and preclinical topics and is now addressing key research areas of special relevance to Singapore such as metabolic diseases, infectious diseases and diseases of the skin.

Another new School has been established within the College of Science and this is the Asian School of Environment with its emphasis on all aspects of environmental sciences especially in the Tropics. There are very few leading world universities located in the tropical zone and all, including NTU, have an obligation to pursue such activities as consciousness increases generally about the need for sustainability. In addition, the College of Humanities, Arts and Social Sciences has expanded and enhanced its activities to become an important all-round College and complementing the rest of the university. Within it the School of Humanities and Social Sciences has become the largest School within the university, larger even than the largest engineering Schools in this technological university.

In terms of the total research community at NTU, the faculty has been reviewed, renewed and reinforced with strong international recruitment of senior and junior faculty. The result is that NTU has become a leading competitor for external research awards, gaining its centres of excellence and several prestigious high-level and very substantial research awards (the so-called 'Tier 3' grants). These much sought-after awards for high-level research projects have been largely led by such senior recruits and cover areas as diverse as disruptive photonics technology, telomere dynamics and genome function, anti-bacterial macro-molecules and, excitingly, the genomic characterisation of urban atmosphere. With faculty numbers today at 1,668 (including NIE), a tripling of PhD students to 3,558 and 2,613 research staff (post-doctoral fellows and research fellows largely funded through external funds), NTU now has a very substantial research presence in Asian academia. A university is vitally dependent on people and this high-level overseas recruitment has been essential as this volume of Singaporean research talent was not available. This is a challenge for the future and NTU, and the other universities, have to ensure that, in the future, there will be a substantial Singaporean presence in the research teams.

The commitment to inter-disciplinarity continues through the pan-university institutes, the Inter-disciplinary Graduate School and the establishment of a new Complexity Institute, to address key problems of urban society and environment.

By strategic planning approaches, NTU has been able to husband and focus its resources to greater effect. As the planning process moves into its second iteration, NTU will focus on five major interdisciplinary areas termed 'Peaks of Excellence': Sustainable Earth, Healthy Society, Global Asia, Secure Community, and Future Learning Paradigm (this latter heavily based on the excellence of the NIE). The Peaks provide a broad inter-disciplinary and coordinating structure involving Research Centres of Excellence, pan-university structures and the expertise of the Colleges and Schools. They represent mature areas of research that, nevertheless, have to stay abreast of advances and renew themselves. The 'Peaks' build on proven excellence in research so that, for example, 'Sustainable Earth' is built around the Nanyang Water Research Institute (NEWRI) and the Energy Research Institute (ERI@N), 'Healthy Society' is centred on the Lee Kong Chian Medical School and centres such as the Institute of Structural Biology, 'Global Asia' on the Nanyang Business School and key centres in the College of Humanities, Arts and Social Sciences, 'Secure Community' on the RSIS and 'Future Learning Paradigm' built around the NIE's substantial research capacity which enables research and the implementation pedagogical innovations and practices.

Not everything can be subsumed under the Peaks, so Focus Clusters have been defined to represent disciplinary and well-established competencies, especially within the Colleges and Schools which may contribute to Peaks or later evolve into new Peaks and are, themselves, areas of significant investment. One example is the creation of a Photonics Institute to establish Singapore as a world-leading centre in photonics technology research. We recognise that research is dynamic and NTU has to prepare itself for new advances in high-level research and be responsive to anticipated changing societal priorities. Emerging and Aspirational Areas meet this need. In terms of our future development one may envisage food sciences and technologies, and architecture being added to the educational and research portfolio.

Of course, all of this has to be based on an underpinning state-of-the-art infrastructure with well-found laboratories and strategic investment in advanced analytical instrumentation and computing facilities. When NTI was created, it occupied the former Nanyang University campus. It was designed by the influential and prize-winning Japanese architect, Kenzo Tange. With the medical school at its downtown Novena campus and with another base at One-North, NTU has now become a multi-campus operation.

At the time of writing, a major building programme is transforming the campus while still remaining true to the original Nantah concept of the tropical Yunnan Garden Campus, acknowledged to be one of the most attractive campuses anywhere. New learning hubs are being constructed putting students at the heart of the learning process together with investment in residential accommodation for students and the general enhancement of campus life. As a residential university NTU will soon have accommodation for 16,000 students, including graduate students as well as housing for around 1,000 faculty. This is a unique asset for the university in education, research and a social and stimulating environment for all concerned.

The reform of the educational programmes has continued and a particular new initiative has been the introduction of a Renaissance Engineering Programme (REP) which attracts top students and provides opportunities for overseas experience through partnerships with University of California, Berkeley, and with Imperial College London. Yet NTU remains true to its original mission to educate and provide practice professionals for the nation through its output of engineers, designers, communicators, accountants and managers, medical doctors and professional researchers and, most importantly, future leaders in all walks of life.

NTU has developed extensive worldwide partnerships on the basis of mutual academic esteem with leading universities in Asia, Europe and North America. These cover collaborative research, joint supervision and joint degrees for PhD students and an enhanced student exchange scheme that aims to provide up to 80% of students with overseas experience. NTU was instrumental in and led the initiation of the Global Alliance of Technological Universities bringing together a network of the world's top technologically-based universities with the aim of addressing global societal issues, for example, biomedicine and healthcare, sustainability and environmental change and changing world demographics, in which science and technology will contribute to their solution. Apart from NTU, the other members are ETH Zürich, Georgia Institute of Technology, Imperial College London, IIT Bombay, Shanghai Jiaotong University, Technical University of Munich, and the University of New South Wales.

Building from an original international recruitment in the previous decade, NTU has fully embraced internationalisation, especially in its recruitment of research staff and faculty as well as research students to the extent that the university now has at least 70 nationalities represented in these categories.

NTU has continued and expanded its tradition of working with industrial partners. Apart from a renewed emphasis and encouragement for innovation by students, researchers and faculty, it has always considered that one of the main benefits for the Singapore economy has been to work with leading technologically-based multinational companies. This helps 'pin' these companies and their investment in Singapore through the provision of highly skilled post-doctoral manpower and by the development of intellectual property that can be rapidly injected into the economy. The portfolio of partners includes leading technology companies such as Rolls Royce, BMW, THALES, Lockheed Martin, Johnson-Matthey, Robert Bosch and many, many more. It also includes Singaporean partners such as ST Electronics, the MRT and Keppel industries. The NRF Corporate Laboratory@University scheme has been an important and, at times, decisive element in moving such collaborations forward. This has enabled NTU to be highly ranked in terms of its industrial collaboration, a tradition which it established at its outset. This is combined with academic excellence and societal relevance to create a recipe for excellence. NTU embraces the philosophy enunciated by the late Lord Porter, former President of the Royal Society, who said: "There are two types of research — applied research and not yet applied research." While linking with industrial partners is very important, NTU has not neglected the need to develop a strong innovation process to encourage and capitalise inventions and discoveries from both staff and students and has established a vibrant Innovation Centre on campus which can act as an 'incubator' for such initiatives.

Linking inter-disciplinarity and research and mentoring has been the establishment of the Institute of Advanced Studies (IAS) which now forms part of NTU's 'Global Dialogues' initiative. IAS is well known for its advanced symposia which have been able to attract world leading scientists to NTU, including many distinguished Nobel Laureates. The presence of such role models is important not only in terms of the insights that they can bring but, perhaps more importantly, the example and encouragement which they give to the next generation of NTU researchers. Of particular significance has been the establishment of a Complexity Institute led by two leading experts in the field. The aim of the Institute is to emphasise trans-disciplinary collaboration in pursuit of understanding the common themes that arise in natural, artificial, and social systems through research and teaching on complexity and complex adaptive systems.

After 24 years of existence as a degree-awarding university, NTU is well on track to fulfil its vision of being a great global university founded on science and technology with the mission of nurturing creative and entrepreneurial leaders through a broad education in diverse disciplines. We have expanded the disciplines that we embrace and more are likely as we become an ever more comprehensive institution but always based on our strong technological background and heritage. This is for the benefit of Singapore, its people and their well-being and for its economy. For the future, it will provide a key element in the future development of Singapore based at the heart of the expansion of Jurong.

Chapter 7

Partnering Multinational Corporations in R&D

Low Teck Seng, Raj Thampuran, and Yeoh Keat Chuan

Multinational corporations or MNCs have been the engine of manufacturing growth ever since the EDB began wooing them to set up operations here. In the late 1980s came the economic restructuring that banked on developing R&D capability to create an additional value proposition for extant MNCs to remain in Singapore while attracting new ones here. It was no longer competition on the basis of lower costs but rather on how much smarter the Singapore economy could be. Making the paradigm shift to a knowledge-based economy would give Singapore competitive advantages because a knowledge-based economy capitalises on people, education and experience rather than on mere price. From Singapore's earliest engagements with the MNCs, it benefitted from close collaboration with them as they were responsive to emerging trends that did not as yet have names. In the late 1960s, the relocation of MNCs to Asia — and the globalisation of the world economy — had been prompted in part by US companies wanting to take advantage of comparative gains from lower costs to be found in global manufacturing. Such lower costs came in part from outsourcing the more peripheral aspects of manufacturing as well as from cheaper labour costs. Singapore's choice of manufacturing for export in place of import substitution manufacturing had tapped into this emerging trend. Looking to lower operational costs, MNCs were breaking up traditionally integrated manufacturing, relocating and using local SMEs to make some of these components. The vertical disintegration of large companies benefitted the emerging industrialising countries such as Singapore. As US and Japanese companies moved manufacturing overseas the production of components came to be outsourced increasingly. MNCs partnered local SMEs and set up supply chains that became increasingly complex and global in nature. Such local SMEs moved up the technology ladder as a consequence, as did the workforce which had to keep up with the new technology. By the end of the 1980s, the process of vertical disintegration, technology transfer and divesting of certain services through the setting up of global supply chains outside the company had come to be termed "outsourcing". This was a trend, now an established practice, identified by management and business school gurus at the end of the 20th century. Except for the most basic, many products today are made up of parts and intellectual property that come from everywhere, sometimes from rivals. Globalised economies emerged, and "globalisation" joined the list of newly-minted economic terms and realities.

Singapore's more highly focused push into R&D as an engine of economic growth after the establishment of NSTB in 1991 could not have come at a better moment. In the 1990s, one of these outsourcing trends surfaced in the area of R&D, one that a business school professor, Dr Henry Chesbrough, Adjunct Professor at University of California, Berkeley, Haas School of Business, gave the catchy label "Open Innovation" in his book *Open Innovation: The New Imperative for Creating and Profiting from Technology,* published in 2003. Dr Chesbrough gave his definition of "open innovation in an article titled *Everything You Need to Know About Open Innovation* in online *Forbes* dated March 2011: "Open innovation is 'the use of purposive inflows and outflows of knowledge to accelerate internal innovation, and expand the markets for external use of innovation, respectively.' Open innovation can be understood as the antithesis of the traditional vertical integration approach where internal R&D activities lead to internally developed products that are then distributed by the firm. As my definition suggests, there are two facets to open innovation. One is the 'outside in' aspect, where external ideas and technologies are brought into the firm's own innovation process. This is the most commonly recognised feature of open innovation. The other, less commonly recognised aspect is the 'inside out' part, where un- and under-utilised ideas and technologies in the firm are allowed to go outside to be incorporated into others' innovation processes." It can be seen from Dr Chesbrough's description of open innovation that since 1991 various aspects of NSTB/A*STAR activities fit this description. Another term to describe what A*STAR has been doing is "knowledge brokering". NSTB/A*STAR evolved Singapore R&D into an open innovation ecosystem using the multi-agency approach. While the EDB was drawing in the MNCs and encouraging them to establish R&D facilities here, the RIs were establishing their research credibility and in incremental steps setting up collaborative research projects.

After 1986, the EDB set up two new units, one of which was the Small Businesses Bureau to grow local enterprises. (See Chapter 8.). The other was the Services Promotion Division to grow the services sector. The strategy called for attracting MNCs to set up their Asian operational HQs here to capitalise on Singapore's excellent facilities as a global transport hub, the country's use of English as the lingua franca, the speed of telecommunications, and its total commitment to commercial enterprise. The huge market opportunity in Asia coupled with Singapore's sterling reputation as the place in Asia to do business and to get business done in a politically stable climate strengthened EDB's arguments to persuade MNCs to set up their headquarters in Singapore. After the formation of NSTB/A*STAR, one of the key HQ activities that EDB worked to bring to Singapore was corporate R&D and to tie up such corporate R&D with the universities, RIs and other agencies to build up knowledge capital. Knowledge creation had become a resource.

Bringing different kinds of entities together to achieve common goals is known as partnering, and it is the trend today. There are even websites and publications devoted to partnering. Whether in technology or life sciences, big companies are in search of external partners to share in building up knowledge capital. In the life sciences, for example, *partneringNEWS*, an online magazine for life sciences executives reported in March 2011 that its partnering conference in Milan had a 63% increase in participants. The same report announced that a 2011 partnering survey conducted by EBD Group that publishes *partneringNEWS* with Thomson Reuters showed that life sciences executives ranked one-to-one meetings at partnering conferences equally as important as their network of personal contacts and as the most significant resource in their partnering activities. The report headlined *Demand for Partnering Increases in High Risk Industry* said: "In the area of alliances, a majority of survey respondents told EBD Group that they expect to see an increase in biotech-to-biotech partnering, while just under 50% said they believe pharma-to-pharma partnering will remain at current levels." The survey of partnering activities was sent to 5,800 executives in biotech and pharmaceutical companies.

On this collaborative trend in R&D versus the traditional internal integrated approach, NRF CEO Prof Low Teck Seng said: "To fund research on their own is very, very expensive. A lot of corporations are moving towards open innovation. Open innovation means they work with anybody and then, when they see something exciting, they'll take it and then further develop it. It's a long process to take something from fundamental research to final exploitation. Along that pathway there are many opportunities for exclusivity. So you do joint work together. When something comes out from it, each company can take the science and develop it into a product for themselves and they can take slightly different paths from the point that it goes into the company. Whether it's in drugs or technology, there is always duplication of some kind. Is a Samsung mobile-phone better than an iPhone? How different is the technology? The technology is all the same or overlapping. Patenting is important to protect your intellectual property. Having established the fact that this is your intellectual property, you can then use it in a collaboration. If a potential partner says 'Let's collaborate', the next question is 'What have you got to offer?' You can say, 'I can offer you ideas, I can offer you my patents. This is my background, my intellectual property which I bring to the table.' Other people will bring their intellectual property. This background intellectual property is brought together and then we do more work and we develop what we call foreground IP, the new intellectual property. In the world today, there are lots of companies that work together in consortia, for example, and they do this for the pre-competitive level. Then they take the science and whatever comes out from the consortium work, they develop them into viable products."

Economists sometimes use the term "gains to trade"; another is "patent pooling" for this process of collaborating to produce innovation. One of the earliest examples of R&D collaboration for a desired outcome must be the pre-war Manhattan Project that developed the atom bombs that were dropped on Hiroshima and Nagasaki. It was made up of exiled European physicists, US, British and Canadian scientists backed with the might of US industrial power. In the 1980s, the human genome could be sequenced so rapidly because it was a collaborative effort of many different scientific groups around the world pooling their knowledge, equipment and research efforts. Open innovation and multi-lateral collaboration make possible consortia like A*STAR's Aerospace Consortium, Electronics Packaging Research Consortium, and Singapore Bioimaging Consortium that bring together multinational RIs, universities and seemingly competing companies to do joint research. Numerous examples of consortia and collaborative partnerships can be found globally in research-intensive countries. But it took more than simply wanting such collaboration and partnerships to bring them about. The RIs and EDB had to work at it. The initial stages were establishing the institutes and building reputation. Once there was reputation to trade on, the search for collaborative partners could begin.

A consortium is one way to engage in open innovation. It is not just a talking platform, says A*STAR Managing Director Dr Raj Thampuran, "We actually do things. You can say 'let me model this technology process'. Everybody is interested in the modelling even though it is not in their core business or not yet in their core business. Their people start to develop those capabilities in being able to model through partnerships in a certain pre-competitive area. It's almost like a teaser. It gets people thinking. After a while, companies are going to say 'Maybe this should go into my technology road map some time soon, and I had better start to do some development work, let's take it bilateral'. In one-on-one projects, companies already know the problems they have and the solutions they want. Consortia are a step upstream where you start to introduce new technologies and then companies start to develop their own insights, some understanding of the new technologies and decide to start their own R&D operations. Then we become the automatic partners because we were the ones to introduce these ideas. By harnessing the synergies of private-public-research collaborations in the pre-competitive space, Singapore is laying the ground for innovation through Singapore-made technologies. By supporting and working with consortium partners, companies in Singapore get a head-start through manpower training and access to the R&D results to prepare them for new technologies and advances. So that's the consortia. And over time we have developed some very sophisticated models. We have got public-private, private-private partnerships. We have evolved over 20 years of building industrial collaboration. If you do a project every two hours, you get to a point where you can design the collaboration structure in ways that are interesting to companies."

Harnessing the Synergies of Collaboration

Even before the 21st century trend of partnering emerged, Singapore's first major public-private sector cross-border research collaboration was put together by the EDB's London office after the Institute of Molecular and Cell Biology (IMCB) became operational in 1987. It arranged a meeting between IMCB advisory board member Dr Louis Lim and Glaxo's research group. At the suggestion of EDB, Dr Lim who was with the Institute of Neurology at the University of London (INL) worked out a research proposal on the Central Nervous System and submitted it to Glaxo. The proposal was for a collaboration between Dr Lim's INL, IMCB and Glaxo. As the research proposal fitted with Glaxo's interest in new drugs for brain diseases, the company responded with a $50-million grant spread over 15 years to fund the research. The researchers rotated between London and Singapore. The collaboration boosted Glaxo's reputation in the biomedical research community and raised IMCB's international profile. The collaboration worked so well that IMCB was sometimes erroneously regarded by the international academic community as Glaxo's institute in Singapore, said Lee Suan Hiang in *Heart Work*. The success of this collaboration encouraged Glaxo to explore other projects. In 1992, Sir Richard Sykes, then President of Glaxo Research, came to Singapore to explore a research project on high throughput for the screening of the active molecules in natural products. Said Teoh Yong Sea in the A*STAR commemorative publication: "(W)e were able to conclude an agreement on the project parameters in one meeting because of the tremendous goodwill that existed among the parties." Glaxo put up $20 million, IMCB did the research and provided infrastructure support worth $10 million, and EDB topped up with $10 million from its National Biotechnology Plan budget. The project was the Centre for Natural Products Research formed in 1992 as a unit within IMCB. Ten years later, the Centre with its invaluable collection of natural plant samples would be spun off as MerLion Pharmaceuticals Pte Ltd. Today the company is focused on developing its lead antibacterial candidate, finafloxacin. Sir Richard who later became chairman of the Glaxo Group and Rector of Imperial College when he retired from Glaxo would come to have a long-term involvement with biomedical science development in Singapore. Since 2000, he has served as chairman of the International Advisory Panel for Biomedical Sciences.[1]

This trend of partnering to improve outcomes and make impact is a strategy actively used by all Singapore agencies. Reputable partners can come from industry, other research institutes, private or public, academia and across borders. In 2011 the Italian FIRC Institute of Molecular Oncology (IFOM) and A*STAR announced the launch of the IFOM-p53Lab Joint Research Laboratory for cancer research in Singapore to be led by an A*STAR scholar, Dr Cheok Chit Fang. The Milan-based private research institute focusing on tumour formation and development is among those who believe in international collaboration for cross-fertilisation of ideas to achieve results. At the time of the announcement of the collaboration, its scientific director Marco Foiani said: "You can't do everything alone. From early diagnosis to targeted therapies, it is clear that if you want to put yourself in the best conditions you need to leverage your resources by finding world class research partners that have similar research interests. The traditional narrow land attitude will ultimately lead to failure: in an increasingly competitive world, establishing long term alliances on projects of common scientific interest is key to achieving first class results which matters not only to Italy but to the entire community. In this framework, A*STAR is an ideal scientific partner because it's a growing scientific reality producing high quality scientific results and their approach to international collaborations perfectly matches ours."

In such partnerships that transcend international datelines, research time is also much increased. In the NSTB era when the engineering RIs were trying to promote themselves to the MNCs in Singapore, they used the "sunrise to

---

[1] Sir Richard was conferred honorary Singapore citizenship in 2004. — http://eresources.nlb.gov.sg/infopedia/articles/SIP_1871_2012-02-20.html. Retrieved on 18 July 2015.

sunrise" argument. Manufacturing problems could be handled in Singapore while US labs were sleeping. When the US woke up, the labs there could take over the work from the Singapore labs. Backing this argument was the beginnings of the Internet, the development of which had been included in the first National Technology Plan in 1991. The National University of Singapore set up Technet in 1992 for the use of the R&D community before it was opened up to the general public in 1994.[2] Technet's connections were mostly with the US universities and research labs. Singapore was among the first handful of countries in Asia, certainly the first in the region, to be connected digitally across borders and to use these connections to further research capability.

Prof Low Teck Seng shares a story about Singapore research and disk drive failures when he was heading Data Storage Institute in the 1990s. "In the early days, when we first started out in disk drive manufacturing, one of the key concerns was failures. Disk drive failures are very costly to the companies because of the warranty. It's very costly if you have to replace it. To produce a disk drive is very cheap, but once it goes out, the whole supply chain that needs to support the delivery of that drive to the consumer is very costly. So this company had a problem. The disk drive has got a head that reads the data, and they found that it was like cheesecake. That means a lot of holes. That means there's something very wrong, so it causes a failure. So they came to see us. At the same time, to give themselves comfort, they sent the whole system back to the US for their labs to do tests. Of course, we solved the problem for them within 24 hours. Once you solve the problem, production can start again. The production delay would have been longer if the failed drives had to be sent back to the US. It takes more than 24 hours for the parts to reach their labs in California from Singapore. By the time the result comes back again, it's four or five days gone. If you need to stop production for four or five days, the cost to a company is enormous. That's one real value of having a research institute like DSI here. So for that reason we were able to continue maintaining a vibrant storage industry in Singapore for more than a generation, for 30 years. Since the end 1990s, hard disk drive companies have also set up capital-intensive and knowledge-intensive hard disk media manufacturing in Singapore. Our success in hard disk drive and media manufacturing seeded an R&D ecosystem. We have Seagate's R&D facility with 500 research scientists and engineers, and Western Digital has established a research, development and design centre here. So we are moving up to very high value-added work. What we have built in DSI

---

[2] Tan, Bernard. The origins of the Internet in Singapore. http://www.physics.nus.edu.sg/~phytanb/bitnet4.htm Retrieved on 12 March 2015.

has allowed us, today, to move rapidly into spintronics and its application in memories and beyond"

EDB frequently encourages MNCs to conduct R&D activities in Singapore, with several examples in the Electronics industry. For instance, Seagate set up a seven-man R&D team in 1984 two years after it started here with a sub-assembly plant for disk drive components. In June 2015, Seagate opened a $100-million R&D Centre (about 500-man) in Ayer Rajah Crescent to focus on developing the technology giant's next generation of 2.5-inch storage drives.[3] Seagate Chairman and CEO Stephen Luczo said at the opening: "For us to push the science where it needs to be pushed, you need to have state-of-the-art labs, facilities and infrastructure. We are fortunate to be in a place like Singapore where we have the human capital." (See Figure 7.1) Another example is the NTU-NXP Smart Mobility Test Bed, a public-private collaboration launched in April 2015. The test-bed for smart cars and traffic systems will validate new technologies in vehicle-to-vehicle and vehicle-to-infrastructure communications. MNCs are not only focusing on developing hardware and software as standalone technologies, but are also developing competencies relating to the integration of hardware and software. As such, Singapore is increasingly seeing investments relating to the development of integrated solutions to solve problems such as those related to healthcare, energy management and security. One example of this is Panasonic's participation in the Punggol Eco-Town test-bed in Singapore to test and commercialise its Total Energy Solution which involves the system integration of several clean energy components such as solar systems, lithium-ion batteries, home energy management systems and energy-efficient air-conditioning.

For the higher-end electronics manufacturing, the A*STAR RIs have played a significant role in helping to attract R&D investments through R&D partnership. For instance, Applied Materials (AMAT), the largest semiconductor equipment manufacturer in the world, has anchored its R&D operations in close collaboration with IME by setting-up a Centre of Excellence in Advanced Packaging in March 2012. This positioned Singapore as the country of choice for global semiconductor R&D. There were about 100 new researchers and engineers positions being created and approximately $90 million in R&D spending has been injected into the Centre to date. The capabilities achieved pave the way for further technology developments such as 2.5D/3D heterogeneous integration on through silicon interposer, chip-on-wafer, fan-out wafer level packaging and emerging technologies such as via-last from backside.

---

[3] The Straits Times, 20 June 2015, p. C7.

Fig. 7.1.   The new Seagate R&D Centre in Ayer Rajah Crescent was named *The Shugart* in memory of Alan Shugart, the founder of Seagate Technology. © 2015 Seagate Technology LLC.

The Centre of Excellence in Advanced Packaging is a powerful combination of cutting edge technology and innovation from AMAT, and industry leading research expertise in advanced packaging from IME. Today, all wafer level packaging research across AMAT is conducted in the Singapore lab and the Centre undertakes complex multidisciplinary research to develop new innovations in semiconductor wafer level packaging (WLP). This has resulted in several multi-million-dollar products in AMAT's development pipeline, the first of which, the Endura® Ventura™ PVD system, was officially launched in June 2014. The joint-lab model also serves as the catalyst to bring further partnership in the broadened value/supply chain. The success of this Centre is an influential factor in attracting similar collaborations with more industry partners to deepen R&D activities in Singapore. In July 2014, another nine industry partners, i.e., Dai Nippon Printing, DISCO, KLA-Tencor, Mentor Graphics, Nikon, Panasonic Factory Solutions Asia Pacific, PINK, Tokyo Electron Ltd, and Tokyo Ohka Kogyo entered into similar collaborations with IME. Four joint labs, focusing on advanced lithography, metrology, wafer level packaging and assembly, resulted from the partnerships. They constitute a combined

R&D investment of $200 million in Singapore. These collaborations which focus on key semiconductor technologies for future growth areas such as 3D chip packaging platforms to achieve high performance, low cost, faster and better consumer and electronics devices, are instrumental in driving industry road map critical for addressing paradigm shifts towards smart era, Internet-of-Things (IOT) and system scaling. In addition, these collaborations encourage increased economic activities as members expand their operations in Singapore, drive local businesses to upgrade their capabilities and create opportunities for the continued development of skilled R&D personnel for a knowledge-based and sustainable economy.

Another industry sector that took shape after the 1985 recession was the Energy and Chemicals sector. Shell's refinery started in 1961 but the development of petroleum, petrochemical and specialty chemicals industries in Singapore accelerated when Petrochemical Corporation of Singapore (PCS) started commercial production on Pulau Ayer Merbau, one of the seven disparate islands in southern Singapore. These seven islands would eventually be amalgamated into a single island known as Jurong Island, which would serve as an offshore location for the energy and chemical multinational companies. The actual work of joining up these islands started in 1995 and was expected to be completed in 2030. However, all agencies involved in the Jurong Island project surpassed themselves by completing the reclamation in 2003. Located on Jurong Island, the Institute of Chemical and Engineering Sciences (ICES) is an autonomous national research institute under A*STAR. It was created to undertake a diverse range of activities from exploratory research to process development, optimisation and problem solving as well as the running of pilot-scale projects. With a focus on providing highly trained R&D manpower, establishing a strong science base and developing technology and infrastructure, ICES was positioned to support energy and chemical companies as they develop new products and processes from Singapore.

Leading industry players across the globe, such as 3M, BASF and Mitsui Chemical have set up R&D facilities in Singapore. In recent years, Singapore leveraged on its strengths as a leading market player in other areas such as marine and offshore engineering, oil additives and lubricants and consumer specialties to grow the chemical industry by focusing on high value-added specialty chemicals. Consumer chemical companies such as Solvay, Croda and Clariant have all sited their innovation centres in Singapore to be in close proximity to their end customers in Asia. In recent years with the setting up of more pharmaceutical, nutritional, and personal care companies, ICES has expanded its research portfolio and collaborations.

Aerospace is Flying High

One of the big success stories is aerospace. The industry's continued growth has been supported by the A*STAR Aerospace Research Consortium which was set up in 2007 to undertake pre-competitive research for the aerospace industry. The Consortium includes major aerospace companies such as Airbus, Boeing, Bombardier, Embraer, GE Aviation, Pratt & Whitney, Rolls-Royce, SAFRAN, local leading companies and agencies such as Defence Science and Technology Agency, ST Aerospace, SIA Engineering, and A*STAR research institutes.

The Consortium focuses on integrating the needs of its members and leveraging on their combined knowledge and experiences to develop solutions to common problems faced in the aerospace industry. This arrangement allows industry members to tap on the different expertise and capabilities of A*STAR research institutes. In turn, problem statements provided by industry members enable the research institutes to stay industry-relevant and perform research addressing real challenges in the industry. Beyond research outcomes, this collaboration platform gives aerospace companies with limited experience operating in Singapore a compelling preview of what Singapore and its R&D ecosystem can offer.

A fitting example can be found in the growth of Rolls-Royce's R&D activities in Singapore after the company joined the Consortium as one of four founding members in 2007. Successful research and technology exploration in Singapore through the Consortium and other bilateral partnerships with A*STAR research institutes encouraged the company's expansion of its Advanced Technology Centre in Singapore. Today, the Advanced Technology Centre has grown to become an important node in the company's global R&D network. It employs more than 50 Rolls-Royce researchers, technologists and engineers to perform in-house research and engineering on manufacturing technology, computational engineering and electrical power and control systems. The fully-established Advanced Technology Centre has continued its engagement and partnerships with A*STAR, NTU and NUS to deliver innovation to its manufacturing and maintenance, repair and overhaul entities in Singapore and around the world. This expanded presence of Rolls-Royce through R&D has rooted the company more deeply on the shores of Singapore, making Singapore a strategic hub for the company in the region. (See Figure 7.2.)

The A*STAR Aerospace Research Consortium has undoubtedly helped to build R&D expertise within Singapore's aerospace industry necessary for its next phase of growth. As Singapore expands its aerospace leadership beyond maintenance, repair and overhaul and in sophisticated after-market services

Fig. 7.2.   A Rolls-Royce engineer in the Advanced Technology Centre performing analysis on an engine component. © 2015 Rolls-Royce plc.

such as fleet management and repair development and manufacturing, the R&D ecosystem co-developed by A*STAR, tertiary institutions and industry will play an increasingly important role in sharpening the aerospace industry's competitive edge over emerging hubs in the region.

Three important factors have spurred the development of open innovation in the 21st century. One is the rapid growth and spread of the latest technology. It has been estimated that 90% of the world's present technology came about in the last 50 years. If we think of the Internet as a web of computer sites with accessible information, the World Wide Web was invented only in 1991 and yet has become a dominant influence in all aspects of life. Another factor has been the rapid growth of innovative technology that has dramatically shortened product cycles. The world's first mobile phone, a Motorola, was unveiled in 1973; the world's first smartphone, IBM's Simon, appeared in 1993. In 2015 the era of the watch phone is here. In the 1960s, it was just a science fiction idea in the Dick Tracy cartoons. The third factor has been the globalisation of technology and the incredible connectivity that has resulted from this. This connectivity has in turn spurred more new ideas, innovations on innovations. Prof Low points out: "To produce anything is not a single technology or a single discipline. There is a lot of fundamental work that is done and is published in conference papers and scientific journals because these findings are contributions to humanity, advancing knowledge. Even top companies contribute to publications and they share knowledge. But from the point that ideas get published to the point where you can develop it into products is not a trial and error process and it's never dependent on only one particular science. They have to integrate with many things together to get a product. Scientists and engineers everywhere contribute bits of knowledge that become realised in innovative, even revolutionary products."

To Boldly Go

The latest area of research that is meriting a much closer look by Singapore agencies is space. In 2011, Singapore launched its first-ever locally built satellite — also the first in Southeast Asia — a result of collaboration between Nanyang Technological University and DSO National Laboratories. Named X-Sat, the 105-kg micro satellite is one of only a small number of regional satellites. In 2013, the Office for Space Technology and Industry (OSTIn) was established with the mandate to develop Singapore's space industry. OSTIn is chaired by EDB chairman Dr Beh Swan Gin, and its steering committee is made up of A*STAR, EDB, NRF and the ministries of Communication and Information, Defence, Education, Foreign Affairs, Trade and Industry, and Transport. Singapore's entry into the space industry is a reflection of the R&D capability that we have built over the years and the increased sophistication of Singapore's adjacent industries such as infocomm, precision engineering, electronics and aerospace industry. For example, National University of Singapore's Centre for Remote Imaging, Sensing and Processing (CRISP) is conducting research in areas like Synthetice Aperature Radar (SAR), and multi- and hyper-spectral data analysis. This expertise is applied in areas such as urban planning, maritime and coastal observation, climate-change studies, disaster monitoring, and agriculture. Major global players involved in satellite remote sensing applications such as DigitalGlobe, GeoEye and Spot Image have established a presence in Singapore to distribute images and manage their regional businesses. Beyond satellites technology and services, sub-orbital flight is looking increasingly possible with advances in technologies. Europe's largest space company, Airbus Defense & Space, has partnered Singapore-based companies and agencies to work towards the demonstration of its sub-orbital spaceplane programme. Space technology is a high-tech niche but it will be one that will capitalise on the capabilities of RIs such as the Data Storage Institute (DSI) and Institute of Microelectronics (IME) which have expertise in memory systems, microelectromechanical systems (MEMS) and power amplifiers. Companies can work with DSI and IME to develop MEMS-based non-volatile memory systems or high-frequency power amplifiers for space. Other research institutes with expertise in application developments in signal processing and data management and analytics include the Institute for Infocomm Research (I²R) as well as the Institute of High Performance Computing (IHPC). Space agencies and companies looking to develop small satellites can also work with Singapore's universities. Both NTU and NUS run micro- and nano-satellites development programmes and whose research areas include formation flying, altitude and propulsion sub-systems and selective critical redundancies development. Singapore even has a home-grown engineering group that is developing as a satellite integrator. ST Electronics (Satellite Systems) Pte Ltd is a unit in ST Engineering that specialises in innovative solutions and services in aerospace, electronics, land systems and marine sectors. Formed in 1997, ST Engineering is now one of Asia's largest defence and engineering groups.

Blooming Biomedical R&D

In linking big and small industry players and research institutes together into various consortia, EDB and A*STAR play the role of integrator or a knowledge broker in facilitating the expansion of knowledge through the sharing of knowledge and stimulating knowledge creation be it on an open platform or in secured R&D facilities. Partnering industry players keeps research institutes more anchored to industrial concerns to realise economic impact while at the same time freeing researchers to explore promising areas of science and technology. EDB and A*STAR work to promote such partnerships to strengthen the funding base for research, make research industry-relevant and to grow the industry sector. There are diverse partnership opportunities with its public-sector research institutes, leading pharmaceutical and biotech companies, clinical-research units in hospitals and international research organisations.

In 2010, Roche announced the setting up of the Roche-Singapore Translational Medicine Hub as its first Translational and Clinical Research site in the world. This 100-million CHF (Swiss francs) ($130 million) investment taps the expertise of local BMS scientists and clinicians from the hospitals, universities and research institutes, and enables them to collaborate with Roche's core team of scientists to conduct multidisciplinary research to accelerate the drug discovery and development process. Also in the same year, GlaxoSmithKline (GSK) established the GSK-Singapore Academic Centre of Excellence. Grants went to researchers at the Singapore Institute for Clinical Sciences (SICS), Duke-NUS Graduate Medical School, NUS School of Medicine, Singapore Eye Research Institute, and the National University Hospital. Its first four projects focused on early-stage research in ophthalmology, regenerative medicine and neuro-degeneration to elucidate new mechanisms of action for innovative medicines. The projects were selected by the GSK-Academic Centre of Excellence Steering Committee made up of GSK's senior leadership and members from the Singapore biomedical research community such as the BMRC, National Medical Research Council, NUS, and Duke-NUS. Bayer Healthcare invested an additional $14.5 million in five projects with local academic institutions to advance R&D to improve early diagnosis and treatment of cancer.

Today, Singapore has established world-class scientific and clinical excellence that enables pre-clinical development and early-phase clinical testing of novel drug candidates to be carried out in one location. In 2011, Maccine Pte Ltd, a pre-clinical contract research company that provides discovery support and regulatory safety assessment services, inked a collaboration with A*STAR's Singapore Bioimaging Consortium to form a comprehensive Translational Imaging Industrial Lab (TIIL) to push the boundaries in state-of-the-art pre-clinical imaging to enhance the drug development process. Siena Biotech, an Italian drug discovery company whose major investor is the non-profit Monte dei Paschi di Siena Foundation, is partnering A*STAR's Experimental Therapeutics Centre to develop molecular inhibitors of a major signalling pathway in oncology to target difficult-to-treat forms of cancer such as gastric cancer, leukaemia and brain tumours. A French biotechnology research and development company, Humalys SAS, and a German public biopharmaceutical company, Cytos Biotechnology, are working with the Singapore Immunology Network to develop antibody therapies for infectious diseases that are prevalent in Asia. While Humalys SAS has strength in its antibody discovery technology, Cytos has focused on the development of targeted immunotherapies with a VLP B-cell vaccines platform, and the Singapore Immunology Network, as with the other A*STAR biomedical RIs, has strengths in certain Asian diseases.

Collaborative research across borders is nothing new, as with researchers from many nationalities. Basel-based Novartis whose primary pharmaceutical research facility, Novartis Institutes of Biomedical Research, is based in Cambridge, Massachusetts, officially opened the Novartis Institute for Tropical Diseases in Singapore in 2002 with more than 100 researchers from 18 nationalities. The Institute plays a role in attracting research talent to Singapore. It has teamed up with A*STAR's RIs and Singapore hospitals, Basel-based Swiss Tropical and Public Health Institute, and US-based The Scripps Research Institute, a non-profit private sector institute, to conduct research. Together, they have discovered a new drug against malaria called spiroindolone NITD609. This breakthrough is timely as drug-resistant strains of malaria are becoming an increasing threat in the region due to irrational drug use and poor-quality or counterfeit medicine. Singapore's growing process R&D capabilities are also helping companies based here to break new ground in both small-molecule and biologics production. A major collaboration between an A*STAR RI and a leading pharmaceutical company is the $2-million public-private partnership between GSK Biologicals and A*STAR's Bioprocessing Technology Institute to collaborate on vaccine and adjuvant system-related research projects.

The number of biopharmaceutical companies has grown from just Beechams and Glaxo in the 1980s. Attracted by the excellent physical and regulatory infrastructure, global connectivity and skilled manpower available in Singapore, many leading biopharmaceutical companies including Abbott, GSK, Lonza, MSD, Novartis, Pfizer and Sanofi-Aventis today make Singapore their global manufacturing base. These companies operate multi-purpose plants with the capability to manufacture a wide range of active pharmaceutical ingredients (APIs), biologics and nutritionals. The country has also made significant inroads in biologics manufacturing, with Baxter, Lonza, GSK and Roche announcing their investments to set up major biologics facilities that amount to $2.8 billion in capital expenditure. All these pharmaceutical manufacturing facilities have been validated by international regulators such as the US Food and Drug Administration and the European Medicines Agency (EMEA).

Many pharmaceutical and biotechnology companies have diversified their operations over the years to include R&D activities besides drug manufacturing. Key examples include GSK grants of close to $8 million in research funding to 14 principal investigators in 2011. This was the second round of projects to be awarded research funding under the $33-million GSK-Singapore Partnership for Green & Sustainable Manufacturing. The principal investigators are conducting research in the areas of chemical-, physical- and bio-transformations, facilities and supply chain, equipment and technical operations, life cycle assessment, and solvent selection and optimisation. The GSK-Singapore Partnership for Green & Sustainable Manufacturing seeks to develop innovative solutions as the global pharmaceutical industry grapple with the need to enhance sustainability and optimise their global operations.

Another project is Lonza's investment of CHF (Swiss francs) 10 million to expand its biopharmaceutical development services platform in Singapore in 2011. Lonza's Development Services business, the front-end of the biological mammalian custom manufacturing process, offers customised and innovative services used in the development of robust biomanufacturing processes. Its expansion in Singapore will consist of the addition of 1,858 m^2 of state-of-the-art laboratory space and associated equipment, and will support cell line construction, upstream and downstream process development, and a broad range of analytical services. The facility will integrate seamlessly with Lonza's biological manufacturing facility in Singapore, creating a full-service biopharmaceutical development and manufacturing site that offers customers a complete range of services from development through pre-clinical and small-scale manufacturing, all the way through to large-scale commercial supply.

Abbot's first major capital investment in Asia and its largest nutritional investment to date is its $450-million nutritional powder manufacturing plant in Singapore. The state-of-the-art facility was completed in 2008 to meet the increased demand for Abbott Nutritional Products in the region. The Singapore facility will produce over 45 million kilograms of powdered nutritional products a year, including milk powders for infants and toddlers. In early 2015, Abbot opened its $32.9-million Nutrition Pilot Plant in Singapore. The facility is Abbott's second R&D hub outside the US and the first in Asia.

After 2010 and the push to incorporate more private sector involvement in biomedical research, BMRC began including notable industry figures on its advisory board. Today, the BMRC board includes directors who are prominent in industry or who have worked for many years in industry. Part of the development of adjacencies that had not been anticipated in the original Biotechnology Plan, the RIs now engage the personal care, food and nutrition industries. In 2011, Procter & Gamble, the world's largest household and personal care MNC invested $250 million to build a mega innovation centre in Singapore, the second of only two such centres in Asia. The 32,000 m², 500-man Singapore Innovation Centre at Biopolis was officially opened in 2014. P&G has also signed a Master Research Collaboration Agreement (MRCA) in 2010 with all 14 of A*STAR's research institutes to allow its Singapore Innovation Centre to leverage on the institutes' research capabilities. In 2013, P&G signed another MRCA with A*STAR that spans five years of research partnerships with Singapore's network of over 25 research institutes, medical institutions and academia.

In 2015, Japanese pharmaceutical company, Takeda Pharmaceutical, relocated its emerging markets headquarters to Biopolis in Singapore. The move will consolidate Takeda's Emerging Markets Business Unit, the Takeda Development Centre Asia and its Vaccine Business Unit into one. Giles Platford, President of Takeda's Emerging Markets Business Unit, described the new office as a "knowledge resource". It will enable the sharing of ideas, marketing insights and practices between business units. He said on the EDB website:[4] "(The new office) will give us close-to-door access to important market information that will accelerate the development of medicines needed for battling regional and global health issues." Among the health issues that Takeda

---

[4]Takeda moves emerging markets headquarters to Singapore by Fiona Liaw. https://www.edb.gov.sg/content/edb/en/news-and-events/news/singapore-business-news/Industry/takeda-moves-emerging-markets-headquarters-to-singapore.html. Retrieved on 20 May 2015.

is focusing on are gastroenteritis and dengue fever. Both are vector-borne diseases, the Aedes mosquito in the case of dengue fever and various bacteria and viruses in the case of gastroenteritis better known to many as stomach flu. Takeda Pharmaceuticals is one of several companies in the race for a dengue fever vaccine of which there are six vaccine candidates in clinical trials by other companies. Other researchers working on this common Asian and South American scourge are scientists in the universities and research institutes. Singapore's Novartis Institute for Tropical Diseases and Beijing's Institute of Microbiology and Epidemiology are in a project to immunise mice with a genetically-mutated, weakened form of the dengue virus as a vaccine candidate. Typical vaccine candidates include live but weakened viruses, inactivated viruses, and portions of the virus' genetic material. At Duke-NUS Graduate Medical School in Singapore, Associate Professor Lok Shee-Mei and her colleagues have found antibodies that are effective against the DEN-1 and DEN-3 serotypes of the virus. Prof Lok is confident of a treatment breakthrough in the next few years.[5] In January 2015, Sanofi Pasteur, a French drug company, announced that it had successfully completed clinical trials in Asia and South America throughout 2014 and was ready to register its dengue vaccine candidate which the company expected to be available by the end of 2015.[6] Sanofi Pasteur was reported to have spent some 20 years researching and developing the dengue fever vaccine which is currently being produced in France.

Also in April 2015, GSK announced the setting up of its Asia headquarters in Singapore. Speaking at the groundbreaking ceremony, Sir Andrew Witty, chief executive officer of the UK-based company, said: "We do not want to have to wait for Britain or America to wake up before we answer a question that needs to be answered in Asia," adding, "And we don't want to send the question back to someone in Britain who has never been to Asia to tell us what to do in Asia." The eight-storey 15,000 m^2 facility located at One-North at Rochester Park is scheduled for completion by the end of 2016.

The pharmaceutical and biotechnology manufacturing sector in Singapore is currently supported by a growing base of more than 4,800 skilled engineers and technicians. Companies can also access a workforce of more than 300,000 skilled employees in related sectors (e.g. chemicals, electronics, engineering). Representing

---

[5] The race to eradicate dengue by Grace Chua. https://www.edb.gov.sg/content/edb/en/news-and-events/news/singapore-business-news/Industry/the-race-to-eradicate-dengue.html. Retrieved on 20 May 2015.

[6] World's first dengue vaccine could be available in 2015 by Denyse Yeo. https://www.edb.gov.sg/content/edb/en/news-and-events/news/singapore-business-news/Industry/worlds-first-dengue-vaccine-could-be-available-in-2015.html. Retrieved on 20 May 2015.

a thriving sector of Singapore's economic growth, the biopharmaceutical sector contributed about $22.8 billion in output and over 6,000 jobs in 2011.

International Reputation and Impact

Global rankings is capital for R&D marketing and they do indicate global perception and recognition of effort. For example, in the 2014 INSEAD Global Innovation Index, Singapore emerged No. 7 out 143 countries.[7] The building of international reputation comes at a price. Since 1991, Singapore has spent billions on R&D. As Prof Hang in a 2012 interview says: "We have raised the R&D spending from 0.96% to about 2% now. We have raised our RSEs per 10,000 labour force from about 30, to now about 100. And we do all this as a percentage of the GDP. The GDP increases, so the actual sum is multiplied by the increase. Those are just figures but what is real is that our international reputation for R&D, serious R&D for industry development has advanced a few notches. We are able to attract top scientific talent here. We are able to meet the manpower requirements of all the multinational companies coming here to set up their R&D centres. They may need 50 or more researchers — we shall be able to supply them with either local candidates or attract foreigners to come. Last year Procter & Gamble wanted to set up their innovation centre in Singapore. They broke ground in the Biopolis, they wanted to hire 400 people. They are now more than halfway. They have hired 200 to 300 people. Not a major issue. If we did not have all these plans to build our reputation, train people and then have the A*STAR research institutes be their partners, we would not have been able to attract these MNC investments in R&D. Without our investments in R&D, even if we had managed to attract them we would not have known how to help them, act as their partners. But now with our investments in R&D, we have the expertise, we have the people, we have the infrastructure. The universities have moved up the global rankings in terms of their research reputation. The number of papers that we publish either in the universities or in the research institutes is respectable. The number of patents that we file is becoming more respectable. So many things have synchronised very neatly. I would say we are on target. We have built this core competence, capability, reputation and ready manpower to contribute to the knowledge economy."

One way to quantify the impact of R&D in Singapore is to look at what EDB has been able to attract to Singapore, the kind of companies, their

---

[7] https://www.globalinnovationindex.org/content.aspx?page=data-analysis. Retrieved on 1 July 2015.

investments, the employment opportunities thus generated, and what they add to the reputation of Singapore as an R&D hub. In the late 1990s, with patient cultivation and imaginative marketing, companies began to set up their research centres here, and the R&D capacity and capability in Singapore was gradually built up. It helped when companies such as Panasonic revealed how their R&D presence transformed their production capability from, for example, designing and producing cassette recorders to DVD recorders, and how, as a result, they not only stayed on despite the lure of lower costs elsewhere, but grew their investments in Singapore with increasingly more sophisticated products.

The indicator of impact that counts for the R&D ecosystem is how many jobs, how many collaborations and how much funding is coming into Singapore from industry partners. Each five-year Research, Innovation and Enterprise Plan (the renamed National Technology Plan) sets a target for industry outcome. The Biomedical Sciences Initiative started off with trying to develop bio-pharma, trying to bring in big investments in pharmaceuticals, investing a lot in drug discovery, upstream science and so on. However, that sector has gone through a consolidation and some dramatic transformations. When the industry-collaboration target for the 2011–2015 Research, Innovation and Enterprise Plan was set in 2010 at $360 million, up from $18 million that was achieved in the previous five-year plan, there was some trepidation about how to meet that target. The 2011–2015 target has been reached and exceeded. For the first nine months of the fiscal year for 2014, A*STAR research institutes brought in over $380 million in more than 1,700 projects with industry partners.[8] According to the 2013 A*STAR annual report: "From 1 Apr 2011 to 31 March 2014, A*STAR has undertaken more than 4,500 industry projects which catalysed more than $600 million in industry R&D investments in Singapore." EDB and A*STAR have also been attracting MNCs in the area of consumer care, food and nutrition. Says A*STAR Managing Director Dr Raj Thampuran: "These are adjacencies because they want a lot of biology and the research findings go into their products. And they are companies that hire a lot of people, they offer quality jobs, and they need skills. So our capabilities now have been applied to the domains that we didn't actually intend in the beginning but discovered the opportunities as they came along."

The diversification into "adjacencies" — the food industry, personal care and cosmetics — all of which call for a deep understanding of biological

---

[8] http://www.businesstimes.com.sg/government-economy/astar-gets-s380m-of-rd-investment-from-industry. Retrieved on 1 July 2015.

science, is not just the downstream stuff. There is deep science in these areas too. Singapore Institute of Clinical Sciences, NUS, NUH and Kandang Kerbau Women's and Children's Hospital are involved in GUSTO (Growing Up in Singapore Towards Healthy Outcomes), one of the world's most comprehensive birth cohort studies involving over 1,200 families, where women were recruited during pregnancy and followed until their infants are at least three years of age. The programme involves and focuses on epigenetics which studies how environmental factors can lead to chemical changes in her baby's DNA, that can in turn affect the child's development and health. This gives us deep insights into early childhood development and the subsequent links to diseases like obesity and diabetes. It also allows A*STAR researchers to work with companies like Nestle, Abbott and Danone, on nutrition products for young children. This deep study involves genetic studies, biomedical studies, collecting of specimens of blood, faeces, urine and so on. As a result, there is now an entire database of what the mother goes through in pregnancy and the effects this has on the baby, the child in the first formative years of his or her life. There have been some very interesting findings. For example, researchers have found conclusive evidence that mothers who had high levels of stress during pregnancy actually had babies whose brains were slightly different from babies whose mothers experienced less stress during pregnancy. MRI scans showed these changes. The findings would substantiate the old wives' tale that a pregnant woman should remain calm and have as few worries as possible to produce a healthy baby. The studies showed the first conclusive evidence that stress is not good for the baby. An even more recent study shows that there may be genetic changes and implications on learning ability. However, translating such findings into clinical practice is difficult. The definition of stress differs from woman to woman. Much easier to translate is the discovery that one in five pregnant women have a form of diabetes during pregnancy. This statistic cuts across ethnic lines, affecting Chinese, Malay and Indian women alike. The discovery came from monitoring the mother's blood sugar levels during pregnancy, hitherto unknown because testing for diabetes in pregnant women was not routine. It has changed the clinical practice so that today, pregnant Asian women in Singapore are monitored routinely for diabetes.

Industry collaboration has not been at the expense of good science because MNCs are not interested in bad science. Says Dr Raj Thampuran: "We have over 1,700 industry projects each year, big and small. Think of almost any multinational and we have a collaboration or some sort of partnership. And the whole idea behind that is to see if we can get them to think about Singapore as a place for internal innovation and develop products here or develop technologies that drive value-added activities here and not just think about Singapore as a business unit or a manufacturing base because those are very sensitive to costs. So the role of A*STAR in trying to interact with MNCs is to catalyse innovation development and not just industry proliferation because that is easy to do, because Puerto Rico offers the same thing. So does Colombia, Malaysia, and Thailand. But if companies start to innovate internally here, it's much harder to displace them. The R&D's value, innovation is value, and that is harder to replicate somewhere else. So when an MNC decides whether to stay or go, whether to stay and grow, the decision it makes is based on how it differentiates Singapore from others. That's the whole issue of competitiveness, how we are perceived, and how we add to it.

"Singapore's perceived value and its strength is that it is a superglue and it is a high-quality adhesiveness. Our ability to bring things together gives us a compactness and advantage that we should not take for granted. It is not just all these things. It is about Singapore Inc, about Singapore being a place where talent would want to come. So then came this whole idea of making Singapore a lot more contemporary in the cultural scene, more accepting of people with different proclivities, and in trying to have aspects in our society that give a quality of life for those whom we attract to come and work and live here wherever they come from. It's our all-embracing way of accepting the best talent from no matter where they come from, the diversity of talent and its porosity. These are not a few things. They are actually a collection of many things that when you bring them together, you have the Singapore Inc story, and that's what I mean by the superglue. It wasn't that only a few parts had the glue, and the other parts happened by luck. It never was luck. It was by design and everything had to come into place. And the time that they came in, even the Esplanade, was actually calculated.

"For the longest time, talent would go where the companies are. If companies go to Switzerland, then talent goes to Switzerland. It goes wherever companies are. At the turn of the century, it has reversed. Today, companies go where the talent is. Talent decides where it wants to live, where it wants to work and where it wants to play, and then companies gravitate towards that. It's a fundamental shift that in public policy when you start to think about creation of jobs, when we could be company-centric in the past, today we have to be talent-centric, and we recognise this. It was actually articulated in the Economic Strategies Committee Report in 2009. And as a consequence we started to put together all these things in place. So the Singapore story why it is attractive is actually this thing, where we saw different parts and then we glued them together. And it's R&D, it's incentives, its people, its infrastructure, it's commitment, it's investment, political stability, English-speaking, schools, One-North. It wasn't by chance. These developments are predicated on a geographic location that is all luck. Singapore is in a region that is developing rapidly. It is the gateway to many parts of Asia, even Australia, to ASEAN, and it sits between India and China, two humongous entities from whom we can benefit by the opportunities that surface from sheer proximity. And Singapore has positioned itself to seize these opportunities by being ahead of the curve in starting to build a knowledge-based economy."

The integrator role that Singapore's R&D ecosystem plays benefits not only MNCs but also local SMEs. It can also benefit external start-ups or foreign SMEs in the knowledge business that hope to leverage on a more vibrant R&D ecosystem to grow their business. Playing the role of integrator, knowledge broker is not new. Since the days of the Science Council, Singapore has taken the lead in the region in matters of science and technology. A*STAR has acted as the National Coordinator and Lead Agency for a number of multi-lateral groups such as the ASEAN Committee on Science and Technology, APEC Working Group on Industrial Science and Technology, International Atomic Energy Agency, and Association of Scientific Cooperation in Asia. Singapore has had lots of practice in bringing researchers together be it for conferences or getting down to brass tacks. Its research institutes have considerable experience in working in consortia and with research agencies, researchers and MNCs from different parts of the world. The trend towards patent pooling, open innovation, and working together is an indication of the growing globalisation of R&D efforts. This move away from the insular and closed approach that characterised much R&D in the 20th century is the future of R&D in Singapore.

Chapter 8

Developing R&D in Local Enterprises

Hang Chang Chieh, Raj Thampuran, and Png Cheong Boon

Local enterprises may be categorised according to economic clusters. There are three main clusters: first, Manufacturing and Engineering comprising Precision Engineering, Electronics, Printing, Packaging, Chemicals, Transport Engineering, Cleantech and Engineering Services; second, Lifestyle comprising Food Manufacturing, F&B Services, Retail, Textile and Fashion, and Furniture; and third, Services and Biomedicals comprising Logistics, Professional Services, Education Services, Healthcare Services, and Biomedicals. Local enterprises may also be divided by size into two groups: larger local enterprises and SMEs. The first group of larger local enterprises comprise companies that may be listed on the Singapore Stock Exchange as well as Singapore-based multinational companies (MNCs). (This chapter deals only with local enterprises. For MNCs see Chapter 7.) Some of the larger local enterprises are technology-intensive companies that may have originated as government-linked companies. Examples in this group would be Singapore Technologies Engineering Ltd, Keppel Offshore and Marine Ltd, and Sembcorp Industries Ltd. Also in this group of larger local companies are those that grew into established companies through effective entrepreneurial leadership. Hyflux Ltd is a well-known example. This group also includes enterprises in the food industry which was one of Singapore's earliest industrialised sectors. Examples are Khong Guan Biscuit Factory (S) Pte Ltd (founded 1947), Yeo Hiap Seng Ltd (now YHS, originated as a soya sauce producer in Fujian Province, China, in 1900) and Amoy Canning Corporation (S) Ltd (incorporated in 1951 in Singapore but started up as a soya sauce producer in Xiamen, China, in 1907). The food industry here was among the earliest sectors to invest in R&D to raise safety and manufacturing processes to international standards, largely through the efforts of the Singapore Institute of Standards and Industrial Research (whose quality and standards components are now under SPRING Singapore). The second group of local enterprises comprise Small and Medium Enterprises (SMEs) numbering 183,400 or 99% of the total of local enterprises (2014 figures). Out of this total of SMEs, 79% consist of micro enterprises with annual sales turnover of up to $1 million, 16% of small enterprises with turnover of $10 million and 5% of medium enterprises with turnover of $100 million.

SMEs are significant employers with seven in 10 workers employed by SMEs, and three in 10 by non-SMEs, and 50% of Singapore's total economic output comes from SMEs. Having a strong SME sector in the economy is very important because SMEs add resilience and diversity to the economy. Apart from being the major employers in the economy, their size and diversity allow them to find niches in different parts of the economy, react faster to changes, thus keeping the economy afloat in the face of changing trends, globalisation and technological disruption. They keep the economy resilient. At the same time, a healthy SME sector reduces dependence on just MNCs which are equally subject to the forces of globalisation and technological change and may even be disrupted in an even more spectacular way than the SMEs. Witness the decline of Nokia once the world leader in mobile phones or Sony once the leader of portable entertainment in the form of the Sony Walkman.

Since the first National Science and Technology Plan unveiled in 1991 and in succeeding plans such as the Research, Innovation and Enterprise 2015 plan, a key government economic strategy has been to develop and grow SMEs through R&D promotion and adoption. The strategy is being implemented by different government agencies. The diversity of SMEs being such, the implementation strategy is not uniform nor can it be a one-size fits all formula. It is in fact highly customised. While some larger SMEs were able or pro-active enough to start or grow their modest R&D programmes with co-funding from the Research and Development Assistance Scheme (RDAS) grants and the tax incentives for spending on R&D with considerable success, the smaller SMEs have required more than monetary incentives to adopt the R&D strategy. Meanwhile, the success of the larger local enterprises with their adoption of the R&D strategy illustrates the importance of R&D in staying economically relevant.

Larger Local Enterprises

Most of the larger local enterprises in Singapore that are technologically driven have found it essential to develop indigenous capabilities beyond what they could license from global technology suppliers. Initially they did not obtain the help of NSTB/A*STAR research institutes (RIs, also sometimes referred to as public research institutes or PRIs) either because their areas of specialisation were so niche that they were not found in the RIs, or their core technologies were developed in-house prior to the establishment of the RIs.

Companies in this category include Creative Technology which became world renowned for their innovative Sound Blaster for the personal computer market and other digital multimedia products. Creative Technology was also one of the first in the world to introduce the digital MP3-based portable music player which disrupted the analogue-based Sony Walkman. Although the Creative MP3 player never became the market leader after Apple introduced the iPod, the iPod's hierarchical user interface owes something to Creative's MP3 player. Having the foresight to file its software matrix patents in the US, Creative Technology sued Apple for infringing its 2001 patents on hierarchical user interface and won an out-of-court settlement of US$100 million plus other favourable commercial arrangements. Today, Creative is driving digital entertainment with cutting edge audio solutions, premium wireless speakers, high performance earphone products, and portable media devices. The company's innovative hardware, proprietary technology, applications and services enable consumers to experience high-quality entertainment anytime, anywhere.

The Real Entrepreneurs

Pioneer EDB chairman Ngiam Tong Dow said: "In the early days, we didn't have any money. So we couldn't put our bets on anybody. So the entrepreneurs in the SMEs, they had to find their own way.... (T)he least promising material, the boys from the Industrial Training Centres, turned out to be the real entrepreneurs.... So now I can tell you, the local companies, our private sector they were started by the boys from the Industrial Training Centres. Some came out and started small factories, some engineers joined MNCs and came out later.

They are the backbone of local manufacturing. The SMEs have come a long way." A shining example of a 1980s technopreneur is Sim Wong Hoo, a 1975 Ngee Ann Polytechnic graduate in electrical and electronic engineering whose start-up, Creative Technology, became the first Singapore technology company to be listed on NASDAQ. Creative Technology began life in 1981 as a computer repair shop where Sim developed an audio card as an add-on for the Apple II computer. That first audio card developed into Sound Blaster technology and Sim's innovation helped transform computers into multimedia devices.

Trek 2000 and the Importance of Patent Protection

Trek 2000 was originally a family-owned electronic components trading business that distributed chips. In 1995, its current chairman and CEO Henn Tan bought the company to provide customised engineering solutions to companies. One day, he was challenged to create a unique product. Tan and his engineers soon came up with a device that could replace the floppy disk — the now-ubiquitous ThumbDrive — the storage medium of choice for millions of computer and laptop users because of its portability and ease of usage. This new creation disrupted the diskettes manufacturing industry.

Without thinking of IP protection, he commissioned a small plant in China to produce prototypes. It was an unwise move given how common piracy of good ideas is in China. By the time he filed the patents in the US, Singapore, and the UK, copycats in China were already flooding the market. Tan was able to get IP protection in the US and Singapore but not in the UK. He also could not afford to file patents globally.

In 2000, Trek 2000 launched the ThumbDrive at the CeBIT international trade fair for information technology and telecommunications solutions in Germany. Response to the product was overwhelming. The success of the ThumbDrive propelled Trek into the global arena.

Leaning on the importance of patents, Trek 2000 today has protected its several solutions around the ThumbDrive concept with some 600 patents. The company's turnover is more than $140 million. Since the advent of the ThumbDrive, the company has won several national and international awards, including one in 2011 from the Institute of Electrical and Electronics Engineers, the world's largest association of engineering professionals.

Trek 2000, another Singapore company with indigenous technology, invented the world's first ThumbDrive in 1998. This storage device soon gained mass adoption globally and killed off the floppy disk. However, being a small local company then and not yet knowledgeable about the process of patent protection, Trek 2000 did not have very strong international patents and missed the opportunity to become the globally dominant producer of the ThumbDrives. From this hard lesson learnt, Trek 2000 has since invested significantly in IP protection for its subsequent inventions, and by 2014 has been granted more than 380 patents worldwide. Its latest innovative product, the Flu Card, has the potential again to revolutionise the data storage market and become an industry standard. The Flu Card is a secure digital memory card that allows photos and videos to be wirelessly transmitted between devices such as cameras and mobile phones. With the advent of cloud technology, the Flu Card could be used to upload photos and videos via even unsecured hot spots for storage in the "cloud". (See Figures 8.1 and 8.2.)

Although R&D assistance was not available initially to these larger local enterprises that were technologically ahead in the late 1980s and early 1990s, many of them eventually gained direct assistance from NSTB/A*STAR and EDB through grants from the Research and Development Assistance Scheme (RDAS) or Research Incentive Scheme for Companies (RISC), or through research collaborations with the RIs and universities. Keppel Offshore & Marine (KOM) is one example of a large local enterprise that put its grants to very good use. Today, it is a global giant with about 40% global market share in the offshore jack-up drilling rig business and with multibillion-dollar annual revenues. In the 1990s, Keppel's Far East Levingston Shipping Ltd (KFELS, today part of Keppel Offshore & Marine) was building offshore drilling rigs with imported jacking systems. As there was only one reliable jacking system manufacturer at that time, KFELS decided to establish an Offshore Technology Development (OTD) pro-gramme to design its own jacking system with an RDAS grant of $1.4 million from NSTB, from 1995 to 1998. The success of this project enabled KFELS to launch its first jacking system in 1997 for its customer Santa Fe.

Fig. 8.1. The original ThumbDrive in year 2000. © 2015 Trek 2000.

Fig. 8.2. The Flu Card. © 2015 Trek 2000.

To further differentiate itself from other competitors, Keppel's Offshore Technology Development (OTD) programme continued to develop a new self-position fixation system under a second RDAS grant of $0.55 million between 1991 and 2001. Its success led to Keppel filing its first patent. Through continued R&D investment and its own technology and designs, KFELS grew to become a dominant global supplier of the B-class rigs. Including the RDAS grants and its own R&D investments, KFELS's total R&D expenditure for the jacking and fixation systems amounted to $8.6 million. The phenomenal commercial outcome from these two systems was way beyond expectations: the total direct revenue generated between 1997 and 2012 exceeded $1 billion; and the total value of the jack-up rigs delivered and under construction exceeded US$11 billion. Dr Foo Kok Seng who was the young engineer tasked with establishing OTD with a handful of R&D engineers and technicians said of OTD's feat: "Our small team was very committed, and we knew that we had to develop a very reliable and cost-competitive system. As we own the design, we could always tweak the design and so we were more flexible to meet customers' needs." KFELS's then Managing Director Choo Chiau Beng reflected in 2014: "We are thankful to EDB and NSTB which supported our first R&D effort in developing the jack-up system and the fixation system. Keppel gained much confidence from this experience and has since continued to do R&D and become a technology and innovation leader. We are so convinced of the need to keep ahead with our indigenous R&D capability that we have recently established a joint corporate research laboratory with NUS." (See Figures 8.3 and 8.4.)

Another Singapore company that has transformed itself into a global company with more than $500 million annual revenue is Hyflux Ltd. Today, it is a leading provider of integrated water management and environmental solutions with operations and projects in Asia, Middle East and Africa. Before becoming the first water treatment company to be listed in the Singapore Stock Exchange, Hyflux was known as Hydrochem Pte Ltd. Founded in 1989, Hydrochem grew rapidly as a system integrator with acquired technology. In 1998, Hydrochem succeeded in its application for an NSTB RDAS grant to develop capabilities in:

- Purification of industrial water as an alternative source to PUB water for industrial uses; and
- Ozonation method for the treatment of dye effluent and refractory organics.

Fig. 8.3. Dr Foo Kok Seng (centre) working on the first jack-up system. © 2015 Keppel FEL.

Fig. 8.4. An example of Keppel's KFELS B Class jack-up rig which has become the industry benchmark. © 2015 Keppel FEL.

The grant of more than $610,000 encouraged and helped Hydrochem to hire three Research Scientists and Engineers (RSEs) and four technicians, and to build an ozonation pilot plant. The findings of the research team enabled the company to further develop water reclamation processes that have become marketable. The Hyflux research focus ties in neatly with the Public Utilities Board (PUB) and National Research Foundation's (NRF) focus on developing national water supply sustainability. Research is ongoing for better methods of water reclamation.

A Singapore company that illustrates the importance of the research-innovate-or-die principle is Biosensors International Group. It is unusual in being an enterprise in the biomedical sector, one in which getting started and scaling up presents different issues from those in engineering. As a two-year old start-up with 20 people in 1992, this medical devices company that was then a contract manufacturer found its sales revenue declining against the backdrop of newer and more innovative devices on the market. It applied and received a $743,600 NSTB grant to develop a new, continuous cardiac output catheter mounted with pressure sensor at catheter tip for continuous measure-ment of cardiac output and dynamic intravascular pressure measurement. The capability developed enabled the company to specialise in developing, manufacturing and licensing technologies for use in interventional cardiology procedures and critical care. Its key product, BioMatrix, is a drug-eluting stent that utilises Biosensors' proprietary technologies. It maintains an R&D team to ensure that it remains relevant in the medical devices market place.

Today Biosensors International Group of companies develops, manufac-tures and markets innovative medical devices, aiming to improve lives through pioneering medical technology that pushes forward the boundaries of innova-tion. It has become a global company with annual revenue of $440 million and a workforce of 1,300 worldwide. Founder and Executive Chairman Lu Yoh-Chie, recalled in 2015: "R&D is a lengthy and costly process. We were brave to invest heavily in R&D as a small company in 1992, as we wanted to create and control our own future. We were grateful to receive the RDAS grant and we appreciated the help of the universities over the years and the conducive R&D environment. As a market leader, we continue to invest 10% of our revenue in R&D in order to keep our products at the forefront of technologies."

Small & Medium Enterprises (SMEs)

With many SMEs originating as family businesses, SMEs everywhere face fairly similar challenges: leadership and succession issues; responses to intense global,

What is an SME?

In 2011, SPRING Singapore re-defined SMEs as businesses with annual sales turnover of not more than $100 million or employing no more than 200 staff. Enterprises qualify as SMEs as long as they satisfy at least one of the two parameters above, regardless of whether they fall under the manufacturing or non-manufacturing sector. Prior to this change, SMEs were defined as enterprises with fixed asset investments of $15 million and below for manufacturing enterprises, and an employment size of 200 and below for non-manufacturing ones.

The new definition reflects the changing economic landscape and diversified profile of businesses where tangible fixed assets may not adequately reflect the size or stage of a company's development. It helps SPRING to better focus and allocate relevant resources, including grants to assist the SMEs. By using sales turnover or employment, Singapore is more aligned with global practices and this facilitates benchmarking. Guidelines on what an SME is vary globally but sales turnover and employment size are common benchmarks internationally. In Singapore, the various assistance programmes for SMEs fix the eligibility criterion of at least 30% local shareholding being held by Singaporeans or Singapore permanent residents.

One of the key reasons for the SME re-definition is to benefit small enterprises in capital- and labour-intensive industries which, because of their high fixed asset investment and employment size, used to fall outside the old definition, thus limiting their access to government incentives for R&D programmes and

(*Continued*)

technological and competitive market changes; lack of strategic and long-term operational planning; and low productivity yields. The SMEs that are best placed to take advantage of R&D strategy are the larger and more profitable SMEs in the manufacturing sector in particular. These are the SMEs that form part of the supply, manufacturing and services chains for MNCs and the larger companies and make up the backbone of the manufacturing sector. Not only would these SMEs benefit the most from the R&D strategy, they also need it the most. They must continually upgrade their core competencies to keep pace with their MNC clients. The higher upstream supporting SMEs are, the more competitive and challenging is their manufacturing environment. In EDB's latest "Advanced Manufacturing" strategy, different manufacturing technologies are to be brought together to create new competitive advantages and capture the opportunities brought about by new business models, for example, in personalised products and by disruptive technologies, such as additive manufacturing.

As discussed in Chapter 2, after 2000, most of the manufacturing SMEs in Singapore could no longer depend on their MNC clients to help them in

(*Continued*)

productivity gains. The 2011 re-definition has been adopted for SME assistance schemes offered by SPRING and IE Singapore. It is also used for all official SME statistics collection, analysis, reporting and benchmarking. The new definition encompasses 99% of total enterprise.

Economists recognise the role that a strong SME sector plays in a vibrant economy. SMEs are a significant engine of growth and as such, SMEs fill niche sectors of the employment and manufacturing markets and some SMEs act as critical innovation links to larger companies and MNCs which are less nimble than SMEs. Being by definition much smaller than the average MNCs, SMEs are more flexible, easier to restructure, and in theory more responsive to rapidly changing technology.

The government agency responsible for enterprise development is SPRING Singapore (Standards, Productivity and Innovation Board Singapore). It said on its website retrieved in 2015: "We work with partners to help enterprises in financing, capability and management development, technology and innovation, and accessing new markets. As the national standards and accreditation body, SPRING develops and promotes an internationally-recognised standards and quality assurance infra-structure that builds trust in Singapore enterprises, products and services, thereby enabling their global competitiveness and facilitating global trade." Another govern-ment agency largely involved in helping SMEs and their products is International Enterprise (IE) Singapore, responsible for promoting Singapore's international trade and the export of Singapore-made goods.

technology transfer. At the same time, these SMES had to stay ahead of the curve of MNC demands and in anticipation of ever advancing technological changes and new capabilities that take time to develop. Thus, when NSTB was re-organised into A*STAR in 2001, one of the aims was to set up a more effec-tive mechanism for technology transfer particularly for the SMEs. Prior to A*STAR, each institute had its own mechanism for technology transfer. Between 2001 and 2002, technology transfer was done under the existing commercial framework which was KRDL Holdings, said Prof Lam Khin Yong, who was made director of KRDL Holdings which had a stable of start-ups from the work of the extant engineering RIs. In 2002, ETPL Investments Pte Ltd was formed with Boon Swan Foo as Executive Chairman. ETPL is the acronym for Exploit Technologies Pte Ltd. Not everyone was enamoured with the word *exploit* but in the spirit of telling it like it is, then A*STAR chairman Philip Yeo decided to go with it. However, ETPL is only one small part of the R&D strategy. More than just an organised pool of exploitable Intellectual Property was required. The next step for pro-actively helping SMEs was soon taken. In February 2003, Singapore launched a multi-agency initiative — "Growing Enterprises through Technology Upgrade" or "GET-Up" in short. GET-Up aims at creating both immediate and long-term benefits for growth-oriented Singapore enterprises and draws together SMEs, RIs and IP. An A*STAR initiative, GET-Up is today the concerted effort of five government agencies: A*STAR, EDB, International Enterprise (IE) Singapore, SPRING Singapore, and Ministry of Education (MOE).

GET-Up for SMEs

The genesis of GET-Up goes back to late 2002 when Prof CC Hang, Executive Deputy Chairman of A*STAR, was tasked by Chairman Philip Yeo to pay special attention to SMEs by heading a new programme which he called Growing Enterprises through Technology Upgrade (GET-Up). GET-Up is targeted at promising local enterprises in the manufacturing sector which see themselves as "Global Enterprises in the Making" to prepare them for the knowledge-based economy. Operationally, the programme would make use of available financial incentives and assistance schemes from EDB, SPRING and IE Singapore, in addition to the technical assistance and manpower transfer schemes from A*STAR's RIs. To put GET-Up into action, a list of companies that might benefit from technology transfer was first drawn up.

In a bold move, Yeo tasked Prof Hang with visiting 500 companies over a one-year period to find out what they needed to stay relevant, and to come up with ways to help them move up the technology ladder. When Prof Hang pointed out that the approval from the Ministry of Trade and Industry had not been obtained yet, he recalled that Yeo replied: "CC, in my life I have never asked for approval. If things did not work out, I would ask for pardon. But I never ask for approval. I thought of what I wanted to do. If it's good for the nation, I went ahead and did it. So this time around, no difference. No need. We just inform them." So on 2 January 2003, Prof Hang began his full year of visiting the shortlisted SMEs. For five days of the week, Mondays to Fridays, he visited two companies a day, meeting the CEO, checking out the company set-up, and assessing whether or how they would benefit from beefed-up technology and which technology. Each night, after the visits he wrote up his reports: Company A could benefit from new technology in X areas, Company B was not ready — end of story.

One immediate scheme that Prof Hang brought to the SMEs during his company visits was called T-Up which involves human talent transfer from the RIs to SMEs. The original idea of T-Up was based on advice which Prof Hang fortunately remembered from Dr Heinz Risenhuber, a German who served on the first NSTB International Advisory Board in the early 1990s. When asked why German SMEs were so capable and able to support the wide range of German MNCs like Siemens (electronics), BMW (cars), Bayer (chemicals) and other big companies, Dr Risenhuber revealed that many years ago, the German Government introduced a scheme to send PhD researchers from its research institutes to work in the SMEs. In the first year of attachment at the SME, the researcher was not paid by the company. In the second year, the company paid half of the researcher's salary. If the company wanted to retain the researcher in the third year, it would then have to pay the researcher's salary in full. But Yeo decided that anything given free in Singapore would not work or was never fully appreciated. There had to be buy-in from the SME to get more commitment. So he modified the German formula into a two-third public subsidy with the company paying one-third for the first two years. Under T-Up, a Research Scientist/Engineer (RSE) from an RI would be seconded to the companies earmarked for technology assistance for two years, with the companies paying only one-third of the RSE's salary.

In the first year of GET-Up, the company-visit team was headed by Prof Hang, who also chaired the GET-Up Steering Committee. On the company side, the CEO and senior managers were expected to be present to ensure meaningful discussions on key challenges and to make key decisions on immediate follow-up actions. By the end of 2003, Prof Hang had visited 437 companies. He was short of the 500 that had been listed because SARS had hit Singapore. So for six weeks, work was suspended because everyone was cutting down on face-to-face meetings. (See Figure 8.5.)

Today, the GET-Up programme is also strongly promoted by SPRING, the agency that knows the SMEs community the best of all the MTI agencies. It is the agency that handles the plethora of SME issues. To make the combined resources of the various agencies available to participating SMES, the following steps are taken. The first step is a company visit for mutual understanding and to determine areas of required assistance. Depending on the stage of development of the SME, one or more of the following unique programmes under GET-Up will be recommended:

- Operational and Technology Assessment (OTA)/Technology Road Mapping (TRM) — later combined as Operational and Technology Road Mapping (OTR);
- Technology for Enterprise Capability Upgrade (T-Up) Secondment Scheme; and
- Technical Advisors (TA).

Fig. 8.5. Launch of GET-Up and the T-Up pioneer secondees in 2003.

The relevant agencies then take follow-up action. The third step is to tailor a package of technical assistance from the RIs together with the financing schemes available to the company. Among the key financing schemes from the public agencies are the Innovation Development Scheme (IDS) and the Local Enterprise Technical Assistance Scheme (LETAS). Efforts are made to explain to companies not yet familiar with these comprehensive schemes and they are encouraged to apply for such financial incentive schemes where relevant. After the first year, a GET-Up senior team continues to visit selected companies to maintain a good momentum of assistance.

The promising Singapore enterprises in the manufacturing sector are mainly in the Electronics, Engineering, Infocomm and Chemicals clusters. These are also the same industrial clusters for which A*STAR's Science and Engineering Research Council (SERC) has built seven public RIs with over 2,000 RSEs (research scientists and engineers) and RTS (research and technical support) and substantial intellectual properties to share. The enterprises targeted for GET-Up assistance are shortlisted by EDB, SPRING and IE Singapore based on their technology upgrading needs and potential.

Figure 8.6 summarises the three component programmes of GET-Up and their expected benefits for participating firms. Each of the three programmes has specific aims and addresses particular aspects of innovation capabilities. The OTR and TA programmes are broad-based in scope and emphasise the

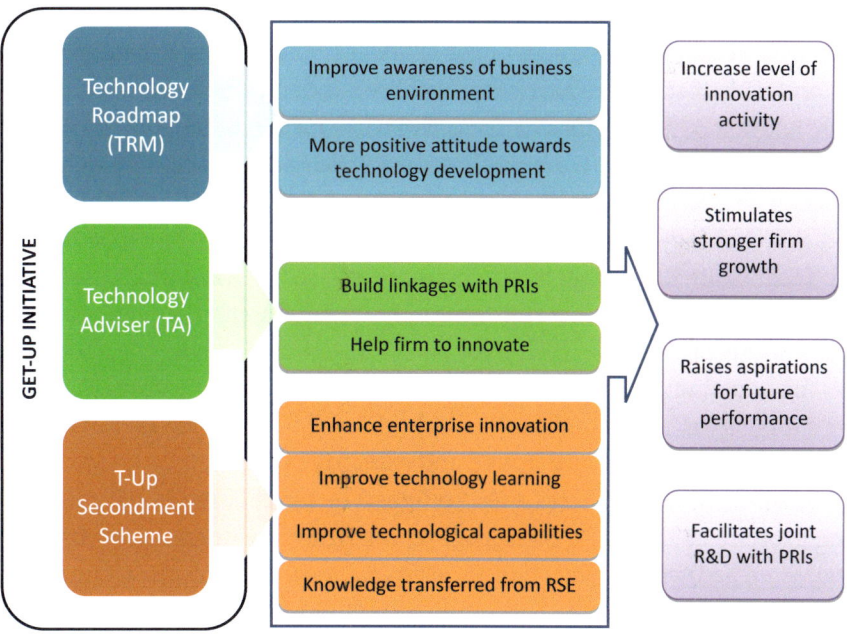

Fig. 8.6. GET-Up programmes and expected impact.

innovation strategies of firms. The T-Up secondment scheme, on the other hand, focuses on well-defined projects. In addition to the successful completion of innovation projects, firms also benefit from the T-Up process, during which they have the opportunity to interact with highly qualified researchers with in-depth domain knowledge in specialised fields.

As a whole, the GET-Up initiative has the long-term objective of improving the competitiveness of enterprises. The strategic and technological competencies imparted by the programme are expected to equip firms with the capability to increase innovation investment and to introduce innovative products and services to market. This should lead to accelerated company growth. Equally important, participation in GET-Up will also promote and inculcate a culture of innovation within SMEs, help them access avenues for cooperation with public sector researchers, and allow them to gain a better understanding of the importance of open innovation in today's industries.

Operational & Technology Road Mapping (OTR)

Because of rapid technological changes and more demanding customer requirements, many SMES need to adapt their business strategies continually to meet new challenges. An Operational and Technology Assessment (OTA) exercise or road mapping is a way of finding out what are the company's operations and technology capabilities to meet future challenges and capture new markets. It is a short two half-day strategic review exercise conducted by experts from one or more relevant A*STAR RIs with the active participation of the company's senior management. When the company has already decided on a definite technology focus, or as a follow-up to the Operational and Technology Assessment exercise, a Technology Road Mapping (TRM) programme may be recommended. The Technology Road Mapping programme is based on a systematic methodology adapted from T-Plan, a fast technology road mapping scheme developed by Cambridge University. Technology road mapping is an effective and flexible tool for organisational planning and for introducing strategic changes. Its relative simplicity is well suited to the needs of small firms which do not have the resources to conduct a complicated and expensive process.

Small firms are often not even equipped to conduct the initial road mapping exercise as they lack experience to recognise the operational challenges involved. The involvement of an RI is crucial as it has both the technological depth and the knowledge of technological trends due to its own long-range research as well as its international networks that include MNC R&D partners and customers. Road mapping is a key management tool that enables companies to link their technological capability to product and business plans so that strategy and technology development will go hand-in-hand. It involves the active participation of the company's senior management over four consecutive workshops led and facilitated by experts from an RI. The workshops, each lasting half a day, consist of:

- Identification of business and market drivers;
- Generation of product/service feature concepts and strategies;
- Identification of technology solution options; and
- Charting of road map linking technology and product developments to market opportunities.

After the completion of the four workshops, the final road map including time frames and milestones will be produced. Senior management in SMEs learn the road mapping process so that they can repeat the workshops by themselves at least once a year to keep pace with the rapidly changing environment. The formal Technology Road Mapping (TRM) programme itself is also very valuable as it helps to create "prepared minds" within the management teams to make sound strategic decisions. The feedback from the companies which have completed TRM under the guidance of the Singapore Institute of Manufacturing

Technology (SIMTech) has been excellent. More companies have expressed an interest in doing both Operational Technology Assessment and Technology Road Mapping programmes. The two programmes together are known as Operational & Technology Road Mapping (OTR) and they have been making good progress in the SME community.

Dr Rob Phaal of Cambridge University, who pioneered the development of this novel planning tool, was invited to Singapore in March 2003 to conduct courses for RI facilitators and also to make an introductory presentation to more than 100 companies. Companies at the presentation were uniformly enthusiastic about the programme. A*STAR RIs have since learnt enough of the programme to conduct such road mapping exercises for themselves. With the momentum generated from the successful pilot run in late 2002, the road mapping programme has since completed 208 plans for 180 companies. In one example of a successful OTR implementation, a precision tooling company found an opportunity to broaden its conventional aerospace industry focus to enter the new biomedical devices market based on its core competence and new capabilities developed with the assistance of SIMTech.

Technology for Enterprises Capability Upgrading (T-Up)

Human talent for R&D is a key for any enterprise in a knowledge-based economy. However, owing to their size and lack of R&D reputation, SMEs that were interested to go into R&D found it difficult to attract highly-qualified RSEs. Being essentially a "brain loan" scheme, T-Up is an invaluable tool in this context. A*STAR's RIs have a sizeable pool of high-quality and experienced RSEs that form part of the T-Up scheme. Under this scheme, experienced RSEs will be seconded from RIs to work in local enterprises for a period of two years to help upgrade their R&D capabilities and create leading-edge technologies that can reap commercial value. The local enterprises are also encouraged to use the two years to court these RSEs and to incorporate them into their own R&D teams. With mutual agreement, the RSEs can be retained by the local enterprises after the secondment period, and the RIs are then able to recruit fresh RSEs to start a new cycle of manpower training. If not, the RSEs then return to the RIs with newly-gained invaluable industrial experience. One important feature of the T-Up scheme is the simple and clear-cut treatment of intellectual property (IP). If certain IPs owned by the RIs are used by the local enterprises, the usual licensing agreement with royalty payment applies. However, if there are new IPs developed by the T-Up seconded staff, the IPs would belong to the local enterprises.

T-Up is the centrepiece of the GET-Up programme, involving the deepest levels of commitment from both public- and private-sector participants. The T-Up secondment scheme aims to upgrade the technological capability of enterprises by:

- Helping to identify critical technologies;
- Building in-house R&D capabilities;
- Forging collaborations between PRIs and enterprises to effect technology transfer;
- Building a culture of innovation and creation; and
- Providing access to human capital and expertise.

Technical Advisors (TA)

It is A*STAR's policy to ensure that there is a constant flow of RSE talent from the PRIs to industry whether it be SMEs through the T-Up scheme or to MNCs. However, the RIs need to retain some of their senior RSEs to serve as master trainers and who cannot be released full-time to the local enterprises. To make the expertise of these senior RSEs available to SMEs, the GET-Up programme has set up a new category of part-time "Technical Advisors" (TAs). Appointed TAs help the CEOs of participating companies in technology planning and strategy formulation over several years but without relinquishing their senior positions in the RIs.

Feedback and Case Studies

The key issue for many SMEs is whether they see technology as the differentiator or can be persuaded to see technology as the differentiator. Much of the work of SPRING Singapore is selling the idea that technology is an important differentiator and that SMEs should incorporate that into their business models. SPRING's experience of the SME community is that when the bottom line turns sluggish, the instinct of many SMEs is to attempt to ramp up sales. For those SMEs who do see technology as the differentiatior, the question then arises whether they know who and what to look for and where they can find relevant information. A programme like GET-Up is the key intermediary that connects SMEs with technology and eventually R&D. Without GET-Up, officers in SPRING joke that getting a researcher to talk to an SME or vice versa is like getting a chicken to talk to a duck. GET-Up acts as a bridge to bring the researchers more downstream while at the same time it brings the SME more upstream so that there is a meeting of minds. Otherwise there is no common area of interest. In fact, many SMEs are so far downstream that SPRING has its work cut out to get these SMES to see that what the RIs do are relevant to them. More often than not, what these SMEs need to see are applications and processes that use the latest IP or the latest technology, to see how a particular research may be applied to their business model. In other words, translation. A good example of such a Eureka moment was when an SME in the personal care business wanted to develop its own face cream and realised that the A*STAR Institute of Bioengineering and Nanotechnology had exactly the IP that they needed. Thus, the company was able to adopt the proprietary technology to produce its own brand name products.

Given that SMEs are defined as companies with fewer than 200 staff, there would be difficulty for an SME setting up its own R&D team, unlike the larger local companies. So it is still about working with R&D partners which means that the R&D partners must also have the ability to reach out to the SMEs. To succeed, GET-Up officers need to understand what are the characteristics of SMEs, what do they need and what they are looking for. SMEs also tend to work on the basis of trust — trust that government agencies do not propose changes that are not good for them in the long term. They are also persuaded by good examples of successes and by the recommendations of people in the same line of business. Thus, part of SPRING's efforts are in generating publicity about the successes of GET-Up and its associated programmes.

As of July 2015, the GET-Up team has visited a total of 2,351 new companies since its initiation in January 2003. Inclusive of revisits, it has visited a total of 3,404 companies. The overall statistics and achievements of GET-Up are shown in Figure 8.7. The general response has been very positive as companies have given feedback indicating that the GET-Up visits were most timely. Most of them were ready for technology upgrading and they welcomed the pro-active approach from the GET-Up administrators. The take-up rate of the T-Up scheme has been good. As of July 2015, there have been 314 approved T-Up applications, involving 569 RSEs. The feedback from the seconded RSEs has also been very encouraging as they have found themselves contributing significantly to the local enterprises. To date, 32 RSEs have been retained by the companies after the secondment period. It has been found from these interviews that one important success factor is that the T-Up candidates can bring tacit knowledge[1]

---

[1]"Unwritten, unspoken, and hidden vast storehouse of knowledge held by practically every normal human being, based on his or her emotions, experiences, insights, intuition, observations and in-ternalized information. Tacit knowledge is integral to the entirety of a person's consciousness, is acquired largely through association with other people, and requires joint or shared activities to be imparted from one to another. Like the submerged part of an iceberg it constitutes the bulk of what one knows, and forms the underlying framework that makes explicit knowledge possible. Concept

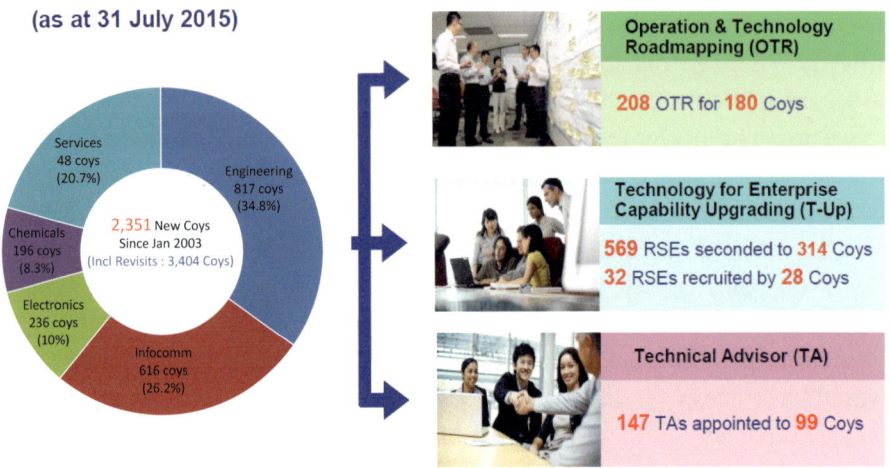

501 Coys had benefited from the GET-Up initiative

Fig. 8.7. Overview of GET-Up achievements. (Reproduced with permission from A*STAR).

essential for further development of technologies or products in the SMEs. This advantage could not be achieved with the normal technology transfer to the SMEs. The very positive feedback from SMEs on T-Up is shown in Figure 8.8 from a 2012/13 survey conducted by NUS.

A good example of T-Up secondment is Advanpack Solutions Pte Ltd (APS) which received the assistance of an RSE, Lim Shao Siong, from the Institute of Microelectronics (IME) to develop Moulded Interconnect

of tacit knowledge was introduced by the Hungarian philosopher-chemist Michael Polanyi (1891–1976) in his 1966 book *The Tacit Dimension*. Also called informal knowledge." — http://www.businessdictionary.com/definition/tacit-knowledge.html#ixzz3ULZnfc38. Retrieved 14 March 2015.

"Codified knowledge can normally be transferred over long distances and across organizational boundaries at low cost.... In contrast with codified knowledge, tacit knowledge refers to knowledge that cannot be easily transferred because it has not been stated in an explicit form. One important kind of tacit knowledge is skills. The skilled person follows rules which are not recognised as such by the person following them. (Polyani, 1958, page 49). Another important kind of tacit knowledge has to do with the implicit but shared beliefs and modes of interpretation which makes intelligent communication possible. (op cit, page 212). According to Polyani, the only way to transfer this kind of knowledge is through a special kind of social interaction similar to apprenticeship relationships. This implies that it cannot be sold and bought in the marketplace, and that its transfer is extremely sensitive to social context." *The Economic Impact of Knowledge*. By Tony Siesfeld, Jacquelyn Cefola, Dale Neef. Published by Butterworth-Heinemann, 1998. page 118. https://books.google.com.sg/books?hl=en&lr=&id=y4SekvWbUr sC&oi=fnd&pg=PA115&dq=tacit+knowledge+definition&ots=I0uI2kYn4w&sig=Ve4RICd43 9Is844_ybp_0-51oVw#v=onepage&q=tacit%20knowledge%20definition&f=false. Retrieved on 14 March 2015.

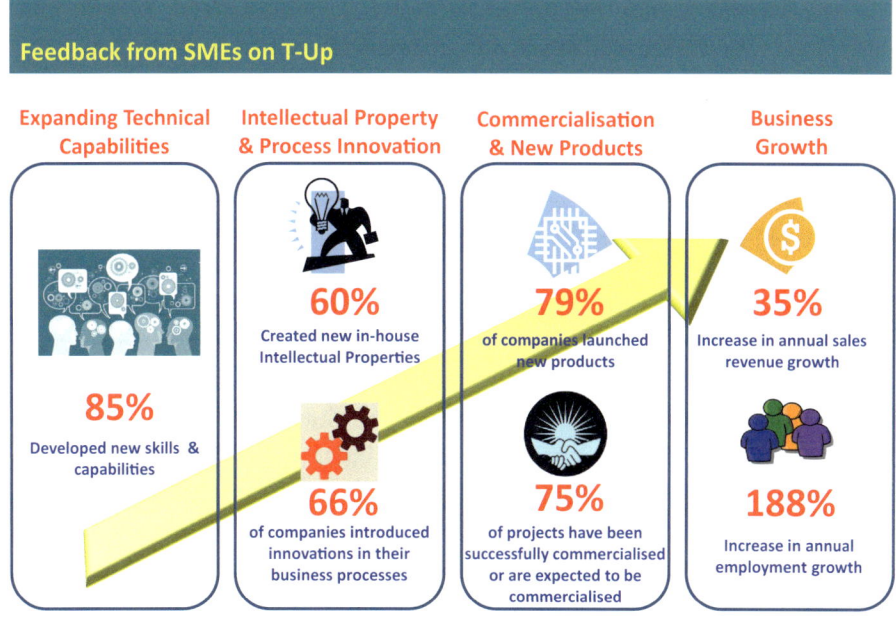

Fig. 8.8. Feedback from SMEs on T-Up success.

System (MIS) and Thermo Compression Bonding (TCB) in 2007. Lim who has a Master's degree in mechanical engineering had been a junior engineer in IME for two years by 2007. When APS approached IME for a possible candidate, he was selected and joined the T-Up scheme. Lim said: "It was a good opportunity to know what was going on in real practice."

Before getting the assistance of an RSE, APS had been focusing on packaging assembling technologies (i.e. copper pillar bump) and processes. These innovations were difficult to commercialise as no one had the capacity to integrate the processes and to meet the affordable price for mass adoption by industry. Hence APS decided to develop its own fully integrated packaging solution to solve this problem. But doing it would require expertise with both upstream and downstream technology know-how. With his background in packaging design and testing analysis, Lim joined the APS R&D team. APS expected Lim to help develop the novel paper lead-frame technology for integrated circuit packaging and to be the liaison person with the research institute to access IME facilities and expertise. It took him half a year to understand what the team was trying to do, navigating different directions with the team, and fully integrated into the company. Reflecting on his learning process, Lim said: "In the two years in IME, I was exposed to a lot of technologies, quite different from what I learnt in university already. Then I got the chance to know even more (in APS)."

In the following one and half years, using his modelling and simulation capabilities, he helped to conduct thermal mechanical stress analysis in the flip-chip design. Knowing the advanced material testing and analysis equipment in IME, he also helped to identify the failures more effectively in the later stage of the packaging design process and hence speeded up the entire R&D progress. After two years, the R&D team successfully developed a new packaging solution which has been awarded one patent and with another six patents pending. The solution has generated tremendous interest and demand and has been licensed to two major clients in the industry. This has supported the company's expansion overseas. Lim also helped to push the R&D in a process innovation which not only assisted the manufacturing of the packaging solution but also generated extra revenue through other applications in the industry.

Besides R&D, Lim has also participated in patent application and IP (intellectual property) management. In 2008, the patent lawyer from Taiwan who had been helping APS to establish its own IP strategy and portfolio left the company and left the unfinished job to Lim's supervisor, CK Ong. Given his previous patent filing experience in IME, Lim was asked to take up the patent filing jobs with Ong. When Ong left the company in 2009, Lim took over the IP management job. He has also helped to link university resources and the company's needs in the area of IP. These new IPs were essential to the company's new manufacturing business model. APS CEO Jimmy Chew said "His work has helped APS to develop its new business model as an IP creation and licensing company."

When asked about Lim's contribution to APS, Finance & Administration Manager Lee Ken Moe said: "We are very pleased with Shao Siong's performance and the ideas he has contributed to our projects. He surpasses our expectations. On top of that, he has excellent work attitude and has adapted very well into our work environment." After two years in the APS GET-Up programme, Lim found that his expertise and interests had expanded by so much that he was no longer interested in the relatively narrow-span research job in IME. At the same time, APS management had found him to be invaluable. Thus, it came up with a premium package to retain him on their payroll. Lim became the first seconded engineer from an RI who chose to stay on with the company. Today, he is the Director of Technology & IP management in

APS, leading two other newly hired IP experts in strengthening and protecting APS's core competency. The packaging solution he helped to develop has been granted 11 patents and licensed to 10 companies.

The APS case shows how the GET-Up secondment scheme can benefit both the SME and the seconded engineer. Not only did the seconded engineer help the SME to develop new products, establish new competitive advantage, increase its revenue and employment, and strenghten its ties with R&D agencies, the secondment helped the engineer to identify his career goals and nurtured his personal development in the industry.

Another example of an SME that made use of an A*STAR RI to grow its bottom-line is JCS-Echigo Pte Ltd. Originally an equipment fabricator, it identified and went into a very niche market of producing cleaning equipment for hard disk drives. It ran into technical problems that the company founder Jason Yeo was then able to solve by turning to SIMTech. The RI connection had been made when JCS-Echigo was visited by a GET-Up team. Today JCS-Echigo is the leading supplier of cleaning equipment for the hard disk drive industry and whose closest rival is an American company. The company has also expanded R&D into other types of cleaning equipment, with the most recent being dishwashing equipment using robotics.

The Technical Advisor programme has experienced a surprisingly strong demand with 147 Advisors appointed to serve 99 companies. These companies can benefit from this new way of tapping RI experts. The first case approved under the Technical Advisor programme was for Tru-Marine Pte Ltd which had requested Dr Lee Loke Chong, a Director of SIMTech, to advise in the area of process technology such as welding, coating and machining. He assisted by acting as a sounding board for the senior management to bounce the company's plans to diversify beyond their core business of turbocharger repair and the feasibility of setting up overseas operations to attend to clients outside Singapore. With access to technology and support from SIMTech and other A*STAR research institutes, the company gained confidence in pursuing these expansion plans.

In another case of how the Technical Adviser programme can contribute to an SME's business, Wan Siew Ping from SIMTech was appointed as the advisor to CEI Contract Manufacturing, a contract manufacturer of printed circuit board assembly (PCBA), box build and equipment manufacturing. Working directly with CEI top management and their staff, Wan helped prepare product development plans, guided the solutions of technical issues, introduced new technologies, raised the engineering competency of procurement personnel, developed strategies for customer and product management, critical staff recruitment, and avoidance of potential problems, among other issues. During her term in CEI, Wan managed to help the company's equipment division to upgrade technologically, and establish a good working relationship between CEI and SIMTech. Even after she was done with her TA term, these well-established relationships continue to benefit both parties.

Many companies have expressed interests in R&D collaborations with RIs as a result of GET-Up participation. The regular interactions and partnership with these enterprises also help the PRIs to better appreciate the future needs of the SMEs and hence to be better prepared to help them. Unlike the T-Up secondment scheme, in this type of open innovation any new IP created in joint R&D projects is to be shared as spelt out in the project agreement. The GET-Up initiative offers participants two avenues for collaborating with RIs on joint R&D projects. Local enterprises with strong marketing strength but little in-house R&D capabilities can outsource some of their product development needs to RIs. A representative example is Stamford Tyres. With its domain knowledge of what exactly is needed for the wheel industry, it teamed up with an RI — SIMTech — to design and develop new types of alloy rims, create the moulds, manufacture them using the latest casting technology, and test them according to current Japanese and American benchmarks. The three-year partnership with Stamford Tyres involved six SIMTech research engineers with expertise in design, moulding, casting, heat treatment and computer simulation.

For local enterprises that have started in-house R&D to meet market demands and have already developed certain core technologies or products, the RIs can help them to further accelerate the growth of their technological capabilities through R&D collaborations. The company can leverage on the larger pool of experts and extensive state-of-the-art equipment of the PRIs to develop and implement leading-edge technologies to capture new and global market opportunities. An excellent example is Singapore Asahi Chemicals and Solders Industries Pte Ltd. Graduating from being a GET-Up participant and beneficiary, the company continued to carry out a number of R&D projects on lead-free solder materials development with the help of SIMTech. The company benefited from the ready access to SIMTech's sophisticated R&D equipment and expertise in structural analysis and materials characterisation. It achieved a breakthrough in a new lead-free solder which is fully compatible with existing manufacturing processes and which met the production needs of a number of MNC customers such as Matsushita and Sony. It intends to continue its R&D collaborations with SIMTech and other PRIs to fulfil its quest for new materials and products for the electronic industries.

The case studies above illustrate how companies have benefited from the GET-Up scheme in different ways. From the perspective of policy-makers, the success of GET-Up contributes to a broader national objective. Commenting on the achievements of GET-Up schemes, Prof Hang, the founding chairman of GET-Up in 2003, said: "The RIs were set up by the Singapore Government to attract and nurture local and foreign R&D talent needed to support its knowledge-based economic development, especially the MNC operations in Singapore. The GET-Up initiative has enabled RIs to help in upgrading Singapore's SMEs which in turn support the innovative operations of MNCs. Among the programmes, the OTR and T-Up secondment were first conceived by A*STAR and then successfully marketed to the SMEs. The TA programme was added because of feedback from the CEOs of larger SMEs. All these benefits and positive experiences have finally led to increased R&D collaborations between RIs and the larger SMEs, which once again has reassured the public funding agency of the strategic importance of public research institutes (PRIs) in the growing knowledge-based economy."

Findings from surveys show that the GET-Up scheme since 2003 has had positive impact on the technological capabilities, innovation performance and growth aspirations of participating companies. Additionally, selected case studies highlight that this integrated approach addresses a wide range of challenges faced by local SMEs and is flexible enough to cater to specific needs and requirements. In return, the PRIs have found a more effective way to fulfil their mission of helping local enterprises by sharing their knowledge and experts through the GET-Up programmes. A better understanding of industry needs has also enabled the PRIs to improve their mission-oriented applied R&D planning and effort.

Instead of traditional technology transfer, the GET-Up scheme utilises human transfer, supplemented by a practical technology road mapping programme, provision of technical advisors when needed, and also joint R&D in selected firms. The pro-active programme by PRIs, with co-funding from public sector grants, has encouraged a large number of Singapore SMEs to participate and succeed in technology upgrading in the face of keen competition in the Asian region. The success of the GET-Up initiative suggests that traditional mechanisms for technological transfer may be refined to emphasise the sharing of knowhow, skills and experience, in addition to transacting technology. To work with MNCs and SMEs, the A*STAR RIs need a wide spectrum of capabilities that range from upstream to downstream. A*STAR's contribution to developing a healthy SME sector is through licensing technologies, making licensing terms friendly for SMEs, creating a wide spectrum of capabilities that will help SMES gear up for growth so they can go any place for whatever technologies they want access to. Through the GET-Up programme, the SMEs get access to equipment, high-level research manpower, new technologies, new product lines, improved productivity, healthy bottom-lines and in the final analysis, continue to be the bedrock of the economy. Because most SMEs are more downstream rather than upstream from the R&D perspective, two more features have been added to GET-Up to make it an even more powerful tool. One programme, Centres of Innovation, taps into the polytechnics' basic and original mission of training for industry. The other is the Technology Adoption Programme (TAP) designed to boost productivity through technology adoption.

Tapping the Polytechnics: Centres of Innovation

Since 2010, the GET-Up Programme has made the industry-oriented polytechnics — Nanyang Polytechnic, Ngee Ann Polytechnic, Singapore Polytechnic, Republic Polytechnic and Temasek Polytechnic — its strategic partners in its effort to help the SMEs. SPRING funds six Centres of Innovation (COIs) in the polytechnics whose mandate is to offer technology consultation and advice at a level that may be more relevant to some of the SMEs than the more upstream expertise offered in the A*STAR RIs. The COIs develop more downstream R&D and innovation capabilities in Electronics, Supply Chain Management, Food Innovation, Environmental and Water Technologies, Marine & Offshore Technology, and Precision Engineering. The feedback from SMEs is that the intellectual property generated by the RIs need to be translated into products or applications that they can recognise as relevant and viable. In other words, the RIs might be too upstream for many of the SMEs who are on the verge of buying into the R&D ecosystem. Thus, the more application-oriented COIs are a better fit for some of these SMEs.

Expertise in the COIs is different from but complementary to that found in the A*STAR RIs. COI researchers may be assigned as T-Up RSEs to work directly with the SME requesting the expertise. If they are not seconded directly to the SME, the expertise is available in the form of consultation services to help SMEs develop their long-term plans to meet the needs of the market. Finally, equipment, facilities and expertise at the COIs are available to SMEs to test-bed projects and applications. This is a mutually beneficial relationship for both the SMEs and the polytechnics. SMEs now have access to an even wider pool of technical personnel to help them increase their productivity and boost their competitiveness. For the polytechnics, participation in the GET-Up programme helps them keep abreast of the latest changes and developments in the industry.

Technology Adoption Programme (TAP)

In addition to the pro-active GET-Up pogramme, A*STAR has introduced a more systematic way for SMEs to access and adopt affordable technology innovations and solutions to raise their productivity. This $51-million Technology Adoption Programme (TAP) was officially launched by A*STAR in November 2013. This Programme was piloted in six sectors: Construction, Food Manufacturing, Precision Engineering, Marine, Aerospace as well as Retail. Supported by a team of experienced intermediaries, SMEs can turn to TAP for advice on on how to improve their productivity. The intermediaries will then link the SMEs with solution providers who are able to develop and implement suitable solutions or technology for them. Where there are no immediate suitable solutions for the SMEs, a group of technology developers in the PRIs and tertiary institutions will help to identify and translate novel technologies for the SMEs to adopt.

By 2015, TAP had reached out to 817 companies and facilitated 110 technology adoptions. It aims to achieve at least 1,150 technology adoptions by companies over three years to increase their productivity levels by an average of 20%. Increasing the productivity of SMEs through technology adoption is a partial answer to the SME calls for ever-increasing numbers of migrant workers to grow their enterprises. Productivity gain is one of the key drivers of wage growth. In a paper prepared for the Ministry of Trade and Industry by economists Tan Di Song and Guo Jiajing titled *Productivity and Wage Growth in Singapore*, the authors concluded: "In this paper, we have shown that while productivity growth is a key driver of real wage growth, real wages may also be affected by other factors such as changes in labour share and relative output prices. In Singapore, the productivity-wage link for resident workers has been relatively strong at the macro level. In recent years, however, falling relative output prices has dampened the translation of productivity growth to real wage growth for residents. This suggests that efforts are required to limit inflation and help the economy restructure.

"At the sectoral level, the productivity-wage relationship is much weaker. We observe different trends for externally-oriented and domestically-oriented sectors. For the former, productivity growth tended to be strong, but the translation to wage gains for resident workers also tended to be dampened by declining relative output prices. By contrast, productivity growth in domesti-cally-oriented sectors tended to be weak, holding back wage growth. Our analy-sis thus suggests that apart from helping externally-oriented sectors restructure and move into higher VA (value-added) product segments, emphasis should also be placed on raising the productivity of domestically-oriented sectors to enable sustainable wage growth in these sectors."

To get wage growth, SMEs must move into economic sectors with higher value-add, and one way to do this is to invest in R&D.

Chapter 9

Towards Innovation & Entrepreneurship

Low Teck Seng, Raj Thampuran, Tan Kai Hoe, and Philip Ong

In its efforts to leverage on research to promote and drive Innovation & Entrepreneurship in both public and private sectors, and in both MNCs and local enterprises, in order to achieve growth, Singapore is beginning to see some successes. Compared to its significant investments in research capability and capacity since the 1990s, the focus on commercialisation and new businesses has only been in the last 10 to 15 years. This can be seen, for example, from the establishment of the Action Community for Entrepreneurship, a public-private partnership to foster entrepreneurship, in 2003; and the renaming of the national "Science & Technology" Masterplan to the "Research, Innovation and Enterprise (RIE)" plan to lend greater focus on innovation and enterprise. For local enterprises, this is a new game. It goes beyond R&D, into design-driven innovation, Open Innovation, Business Model Innovation, etc. The commercial successes of a few large local enterprises and the timely upgrading of high-growth SMEs through the GET-Up Programme of A*STAR and SPRING Singapore (as elaborated in Chapter 8) give hope that with further persistence and entrepreneurial skills, more Singapore SMEs will grow by launching more innovations in the near future. University and polytechnic education will need to be enhanced to provide training and hands-on experiences for engineering students to learn design-thinking, create viable business models, and develop awareness of technology commercialisation, techno-entrepreneurship, open innovation strategies, etc. Entrepreneurs like Sim Wong Hoo and Henn Tan are examples of these possibilities. There is evidence that with pro-active and substantial nurturing of talent and ideas more such entrepreneurs and companies can develop. Many more university and RI spin-off companies, fuelled by recently introduced POC/POV grants, nurtured through Incubation Centres or Accelerators, funded and assisted by a more sophisticated venture ecosystem, will generate a new momentum for innovation and entrepreneurship. New public policies to facilitate and accelerate this are being conceived and will be seen in the latest RIE2020 Plan announced in early 2016.

Innovation and entrepreneurship is the engine of economic growth and economic sustainability. It generates new jobs and contributes to economic renewal when the economy might otherwise be flat. Innovation and entrepreneurship is not only risky and hard work, it calls for creativity and imagination, attention to details, business savvy and funding. Depending on the field of research, there are many obstacles on the road to commercial outcomes ranging from the cultural to the practical. In the field of engineering, commercial outcomes build on other patents or are innovated from combining old and new patents. New products come from innovative ways of looking at familiar things. In the biomedical sciences, on the other hand, research findings must be translated into treatments for disease, into some aspect of healthcare and then be tested clinically. While success is always a tremendous achievement because it contributes to better healthcare, it is also one that requires the exercise of much patience, long hours, constant nurturing, and the ability to handle failure time and time again. Small businesses can become big businesses and even MNCs if the business model is viable and relevant. They may grow not only in the traditional way by building market share and getting listed on the stock exchange but also through the process of mergers and acquisitions which are today part and parcel of growth. Innovators of all varieties are also needed — from those who leave large corporations to start their own business, to those who stay back to be "corporate entrepreneurs", to young students eager to create their first start-ups. These innovators need to be supported by business angels, venture capitalists, and even corporate venture firms owned by the MNCs.

Linked to innovation and entrepreneurship today is the strategy of developing a knowledge-based economy, thus realising economic value for Singapore's investments in R&D. While publication of research findings does much to build up research credibility and illustrate capability, publication per se has little impact beyond extending the boundaries of knowledge and stimulating more research. To get economic impact there must be an innovative and entrepreneurial ecosystem that is an important part of the R&D landscape, an ecosystem that has to include a strong spirit of enterprise. R&D and enterprise have produced revolutionary medical care, drugs that treat once-incurable diseases and extend human life. They have brought about sweeping changes to every facet of life. The recognition of the critical role of innovation and entrepreneurship as an engine of economic growth and an essential part of the R&D ecosystem is seen in the steps taken at a national level to promote the development of innovation and entrepreneurship coupled with research.

The National Level

Underlining the importance of enterprise in the R&D ecosystem is the creation of an innovation and enterprise council on which members of the Singapore Cabinet sit. Set up in 2006, the highest level agency to promote innovation and enterprise is the Research, Innovation and Enterprise Council (RIEC) chaired by the Prime Minister and on which nine ministers sit. It also has an interesting blend of top local and foreign academicians and industry leaders. The RIEC is a policy-making body that decides the directions that Singapore should take vis a vis research, innovation and enterprise. Supporting RIEC is the National Research Foundation (NRF), whose board is chaired by the Deputy Prime Minister and comprises ministers of agencies engaged in research and innovation. The NRF is a department within the Prime Minister's Office, and is the executing agency for the RIEC as well as its secretariat. All agencies involved in R&D draw up plans for the next five-year budget and these plans are tabled to the NRF board, which decides how to manage resources for the next five years to best meet the agenda set by the RIEC. This system gives flexibility for quicker responses to changing economic circumstances but at the same time capitalises not only on the leadership of the ministers but also benefits from the inputs of ministers whose ministries will be impacted. Getting buy-in is important because the nature of science and technology today is such that all layers of society feel the effects of innovations eventually.

The NRF manages the National Framework for Innovation and Enterprise (NFIE), a national programme as shown in Figure 9.1 to grow innovation and entrepreneurship in Singapore. This programme encourages universities and polytechnics to translate their research into technology-based start-ups and into commercial products for the market. The framework comprises a number of schemes to help realise these goals. One is the Early Stage Venture Fund where the NRF invests $10 million on a 1:1 matching basis to seed several venture capital (VC) funds that invest in Singapore-based early stage high-tech companies. Another scheme is the Proof-of-Concept (POC) Grants that give out grants of up to $250,000 for technology proof-of-concept projects. NRF administers the scheme for university and polytechnic researchers while SPRING Singapore runs a parallel scheme for SMEs. There is also the Technology Incubation Scheme where the NRF co-invests up to 85% (capped at $500,000) in Singapore-based start-up companies incubated by seeded

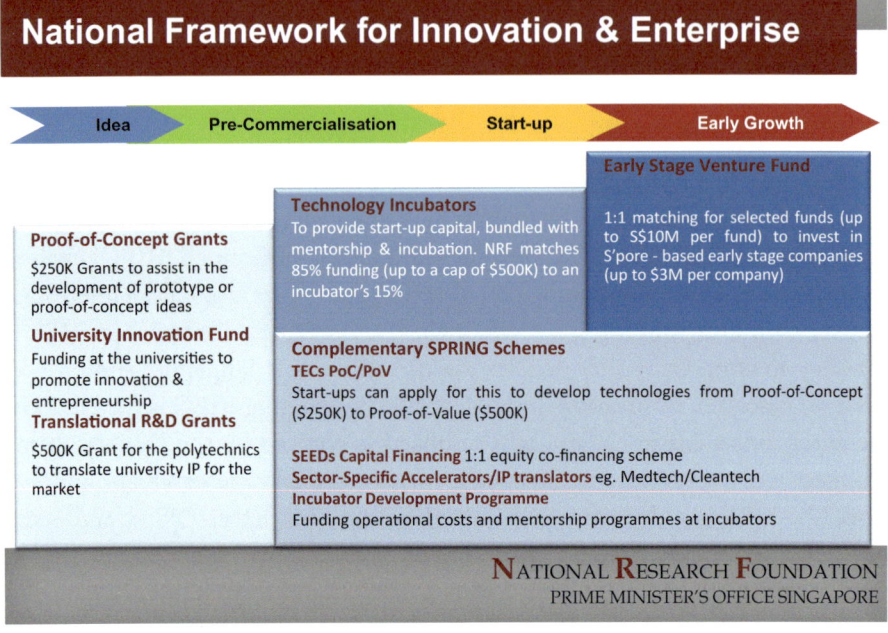

Fig. 9.1. National framework for innovation. (Reproduced with permission from National Research Foundation).

technology incubators that provide their investee companies with physical space, mentorship and guidance.

Two more schemes are the Global Entrepreneur Executives and the Innovation Cluster Programme. The Global Entrepreneur Executives is a co-investment scheme to attract high-growth and high-tech venture-backed companies with global entrepreneurial management in information and communications technology, medical technology and clean technology to re-locate to Singapore. NRF invests up to $4.2 million in matching funds to eligible companies in the form of convertible notes. The Innovation Cluster Programme as shown in Figure 9.2 encourages technology organisations and economic agencies to work with industry to form innovation clusters. It seeks to strengthen partnerships across companies, universities, research institutes and government to bring technology ideas quickly to market, raise productivity, create jobs and grow the sector.

Among the government agencies involved with developing innovation and entrepreneurship in general are SPRING Singapore and International Enterprise (IE) Singapore, and the oldest development agency, the Economic

Innovation Cluster Programme

- Encourages technology organisations and economic agencies to work with industry to develop and execute innovation cluster development plans

- Seeks to strengthen partnerships across companies, universities, research institutes and government to bring ideas quickly to market, raise productivity, create jobs and grow the sector

NATIONAL RESEARCH FOUNDATION
PRIME MINISTER'S OFFICE SINGAPORE

Fig. 9.2.   Innovation cluster programmes. (Reproduced with permission from National Research Foundation).

Development Board (EDB). The main task of IE Singapore is the external promotion of Singapore companies. Its role is to drive the external growth of Singapore-based companies through international trade. Thus, it promotes Singapore-made products and services on the international markets and in this way spurs the growth of Singapore companies.

The role of SPRING is to nurture existing SMEs as well as brand-new start-ups. It employs four strategies, the first and most basic of which is to strengthen the environment for business development and growth through a right blend of policies and funding. The second is to identify development thrusts for specific clusters. The third and fourth strategies narrow the focus onto growth-oriented enterprises. One group is the fast-growing medium-size enterprises and the aim is to grow them into larger enterprises. The other group of growth-oriented enterprises are the start-ups and the strategy is to identify the innovative ones and to grow this sector. Essentially, the strategies boil down to developing different types of capability for SMEs. Other than helping individual SMES to do that, SPRING also looks at the entire systems level — to see how it can catalyse different stakeholders and players to strengthen the whole system. SPRING sees its role in innovation and enterprise as linking the SMEs into the R&D ecosystem to realise impact from these investments. In other words, to see how to translate research into successful products, services or enterprises, both into younger companies as well as to existing companies. In terms of bringing R&D further downstream into actual products and services, its partners here are the Centres of Innovation set up in the different polytechnics but also the public research institutes. At one end is the translation-type work, and at the other more upstream end, the developmental needs of the companies. The coverage is diverse and ranges from materials research to precision engineering to food and nutrition and healthcare among others.

A*STAR advances innovation through active partnerships with companies of different types. These collaborations enable the transfer of technologies across a wide span of areas and the expertise from top talent. Arising from this is the ability to increase the absorptive capacity of companies in Singapore to innovate because of the opportunities to co-create, co-develop and collaborate with research institutes. The Agency takes a differentiated approach depending on the type of company. MNCs and the larger local enterprises typically require customised, integrated solutions that necessarily marshal a variety of expertise to develop new products and improve or introduce new processes. SMEs often require closer-to-market solutions and consequently A*STAR researchers must be able to move further downstream in a diverse range of disciplines which encompasses the SME sector. Another means of technology transfer is through licensing and of the more than 200 licenses each year, 55% are absorbed by SMEs. Today, A*STAR has collaborated with over 80% of the companies that contribute 80% of the private sector R&D expenditure annually and it undertakes over 1,700 industry projects a year.

Another strategy to create more drive towards Innovation and Entrepreneurship is the Intellectual Property Intermediary (IPI).[1] Established under the Ministry of Trade and Industry, the mission of IPI is to promote innovation and entrepreneurship by linking SMEs with the necessary intellectual property that will help them to grow regardless of where that IP may have been generated. Its website says: "If you are looking to harness innovation to stay ahead in a rapidly evolving market, then you've come to the right place. IPI is a catalyst, enabler & partner that connects Singapore-based businesses with solution providers to enhance innovation capability. … IPI's team comprises technology specialists with expertise across a wide spectrum of industries and skillsets to help companies tap expertise beyond their four walls. Our goal is to assess

---

[1] https://www.ipi-singapore.org/about-ipi. Retrieved on 18 July 2015.

your business, help fill the gaps in your technology road map, uncover great innovation opportunities, and bring your business to the next level. Over the years, IPI has developed a global network of technology partners and a technology marketplace featuring innovative technology from a wide range of industries." According to its website, IPI has successfully connected a variety of industry players to solution providers. For instance, Matex International, a home-grown specialty chemicals company, has benefited from IPI. "From inception in 1989, we have recognised the need to develop sustainable, low-carbon impact and effective dye and chemical solutions to vastly reduce the negative impacts on environment as textile dyeing processes are a major cause of industrial fresh water pollution around the world," said Dr Alex Tan, founding CEO & Managing Director of Matex. "Nowadays, the world is so connected and innovation happens at a much faster pace. It is not enough to rely solely on the company's own R&D activities to develop innovations to reduce the time-to-market and business risk, said Dr Tan. Matex currently has multiple projects with external collaborators in the pipeline to develop eco-friendly products and processes. For example, through the technology sourcing and assessment assistance of IPI, Matex was matched with a start-up company to develop and test wastewater treatment technologies. "IPI's facilitation and crowdsourcing solutions have helped the company unlock new capabilities which can create multiple and rewarding business opportunities for Matex to grow," says Dr Tan. IPI also taps into the global IP market, thus increasing the chances of finding a match between SMEs looking for solutions and solution providers, and speeding up the process of innovation and entrepreneurship.

Where Do Innovators and Entrepreneurs Come From?

While enterprise and entrepreneurship are related terms, they are not identical in meaning. An entrepreneur is an enterprising person prepared to take financial risk to achieve economic goals of profitability and growth. Enterprise may refer to a system of business, e.g. "national enterprise", or mean something that someone would like to attempt or do, especially something risky. An entrepreneur must have certain qualities: persistence and the ability to handle failure, imagination and creativity, self-confidence and passion for his ideas, self-discipline to put in the necessary hard work, and a strong competitive streak to want to succeed. Innovators have similar qualities as entrepreneurs except that they may not be involved with an additional personal financial risk. Today, identifying individuals with such innovator or entrepreneurial qualities has come to be critical given the role that innovation and entrepreneurship plays as an engine of growth.

In Singapore the nature of entrepreneurship has changed. In the past, entrepreneurs were people who went into business because opportunities to go into less risky lines of work were non-existent. In the 1950s and 1960s it was the "entrepreneurship of survival". Today entrepreneurship is more nuanced and motivations more diverse. One group of entrepreneurs sees enterprise as a more attractive option than holding down a steady and possibly boring job. This group may go into enterprise armed with plenty of paper qualifications such as MBAs, engineering and medical degrees, and have a good network of supporters and investors to initiate the start-up. Included in this group may be the children of the pioneer entrepreneurs who see more economic potential in going into the family business than getting a steady non-risky job. They may even see the potential of using the family business as a base for striking out on their own into a new enterprise. Some of these new enterprises that grow out from an established position of wealth are the social enterprises that seek to better lives among the less fortunate, promote a more sustainable lifestyle and so on. Another group of entrepreneurs is made up of people who have been in steady jobs and who now want something more challenging, or from their work experience have identified an unexploited niche in the market place that they want to capitalise on. This group of entrepreneurs usually comes with industry knowledge and management expertise. Like the entrepreneurs with paper qualifications, they are well-placed to appreciate and take advantage of the public research institutes as well as the diverse range of incentives and nurturing schemes for entrepreneurs. Then there are the corporate entrepreneurs who also come equipped with industry and management expertise. These are the forward-thinking leaders in established businesses that look out for new enterprises and new business models and exploit new technologies to create new markets in order to survive and prosper. Corporate entrepreneurship enables the companies to compete with other incumbents, fend off attacks by new entrants which are most likely to be start-ups investing in new technology and potentially disruptive ideas. In addition, corporate entrepreneurs may provide funding for some of these start-ups or collaborate with some others through co-innovation, thus opening up yet another opportunity for Singapore.

Another obvious source of entrepreneurship is the community of research scientists and engineers (RSEs), the men and women who do the research and who may often know better than most, whether there is economic potential in their findings. RSEs do not lack persistence and are not afraid of failure. Every researcher worth his salt encounters his share of failures. After all, research is about checking out hunches using scientific methods, not all of which pan out. But it does indicate something when some researchers say that being a researcher is more of a lifestyle choice than a career option. Many are not ambitious enough to take all the gruelling steps necessary to capitalise on a "Eureka moment" provided there is even one that can be capitalised on. Researchers say that such Eureka moments are not that common and in the field of the biomedical sciences, commercialising such discoveries takes many years of further research, translational and clinical trials.

Would-be entrepreneurs must be creative to spot ideas with business potential if they do not come up with their own ideas. They must understand their markets to spot niches that they can exploit. One of the questions raised about developing entrepreneurship in Singapore is whether enough young people can see the big picture as well as the micro features where new business opportunities lie. A standard criticism is that too many young Singaporeans are too comfortable to venture into enterprise. Another criticism is that Singapore's education system kills creativity. Prof CC Hang disagrees. He thinks that while it may block creativity or delay its expression, it is never killed because creativity is the other side of the brain. It is always there and the question is how to draw it out and nurture it. Many examples can be found when Singapore students go to the US for further studies. Most excel in both studies and subsequent careers in the US which equip them to be creative. Indeed they also have the self-discipline to work at it, a self-discipline built by the Singapore education system. To bring out their creativity, it would be advantageous to give them early exposure. Such exposure can never be too early because Singapore with its small cohorts of young people cannot afford wastage. In the end, people cannot be forced to be entrepreneurial though many incentive schemes are out there. However good the funding schemes, all entrepreneurs must bear quite big personal risks from lost opportunity costs and savings and time invested in the enterprise. Spin-offs happen because certain individuals feel very strongly about their ideas and want to see them in the market place. What agencies like A*STAR and EDB can do is to enable such individuals to realise their visions through effective policies, timely funding, research support, marketing support. Research institutes, agencies and companies can create policies that make it possible for individuals to take such risks, reduce the risks and their concerns and uncertainties of going out there and competing in the market, knowing that most enterprises will fail. There are other types of incentive policies that help to encourage, try to incubate technologies as far downstream as possible so that entrepreneurs don't go into the market place too soon. The Singapore R&D ecosystem must think about and create conduciveness to entrepreneurship.

Incubators and Accelerators

A well-established method of nurturing start-ups is through incubators. An incubator shields start-ups by absorbing some of the risks of setting up a viable business such as rising rentals through the provision of space, management advice and marketing know-how. Incubators are usually non-profit and government-funded. By reducing the risks of failure, an incubator gives an enterprise the time to grow and develop its product or service and find viable market share. SPRING characterises its start-up development work as that of an architect and builder by identifying what are the enablers that the start-ups and entrepreneurs need. It then tries to design something that works in a system way. If a private sector catalyst is needed, SPRING then seeks out the necessary partner for the start-up. It tries to bring in large companies, even government agencies, and where possible buyers and test-bedding partners for these start-ups to validate their technology and show its market potential. Adoption of the start-up's technology by an established company or government agency creates a reference customer for the start-up to build its market. (See Figure 9.3)

Fig. 9.3. Incubator development programme. (Reproduced with permission from SPRING Singapore).

Block 71 is one of the latest efforts to give start-ups a leg-up. Started in 2011 by NUS Enterprise, Media Development Authority (MDA) and SingTel Innov8 Pte Ltd, with the support of JTC and SPRING, Block 71 was a strategic incubation programme for start-ups primarily in the IT space. Converted from a disused building in the Ayer Rajah industrial estate, Block 71 addresses the issue of space for budding entrepreneurs to turn their tech ideas into viable businesses. Block 71 also aggregates entrepreneurs to create a collaborative and synergistic community. It houses not only entrepreneurs but also venture funds and incubator/accelerator programmes. It brings together the institutions, industry experts, funding, as well as the government agencies. To help the start-ups connect the dots, incubators like NUS Enterprise regularly hold business clinics, networking sessions, venture capital (VC) pitching sessions, and industrial sharing seminars for the occupants. The start-ups thus get exposure in a holistic support network. Support funding for these start-ups comes from VC funds such as SingTel Innov8, a wholly-owned Singapore Telecommunications Group venture fund for IT start-ups, Vertex Venture Holdings, a wholly-owned Temasek Holdings venture fund, Infocomm Investments Pte Ltd, a wholly-owned Infocomm Development Authority of Singapore venture fund, and SPRING SEEDS Capital, the investment vehicle of SPRING Singapore.

The location of Block 71 is close to A*STAR, NUS, as well as the other ongoing developments in the One-North area. Block 71 has evolved into a space that houses an increasing number of successful local icons of IT entrepreneurship. With this initial success, JTC has added two new blocks, 73 and 79, to provide the capacity to incubate up to 750 start-ups, creating what is now called the JTC LaunchPad @ One-North. At the same time, the partners in Block 71 have exported the model to San Francisco calling it "Block 71 San Francisco". The decision to set up Block 71 San Francisco arose in part from the interest shown by US start-ups in the Southeast Asian market. An Infocomm Investments Pte Ltd press release in January 2015 says: "Infocomm Investments Pte Ltd (IIPL), NUS Enterprise and SingTel Innov8 have together set up Block 71 San Francisco, a US-based co-working space to strengthen ties between start-up ecosystems in Singapore and the US. Singapore tech companies exploring business opportunities in the US can leverage this facility to better understand the US market, 'set up shop' and expand their network into the tech community."

Other than the more traditional incubator model, there are also venture accelerators that provide intensive and extensive support in all areas of start-up growth for a short period of time, with the aim of accelerating the start-ups' success or failure. Also called a seed accelerator, the programme brings together start-ups for a short period of intense talking, mentoring and bouncing off ideas on how to turn ideas into viable businesses. The intense sessions culminate in a Demo Day at which the founders present their start-ups to an audience of invited investors. The seed accelerator model is a quintessential example of enterprise in a knowledge-based economy. It sells knowledge in a way that will generate more knowledge. Helping start-ups to start up was a business model that Paul Graham and his then girlfriend, now wife, Jessica Livingston, Trevor Blackwell and Robert Tappan Morris came up with. Created in 2005 in Cambridge, Massachusetts, Y Combinator is today the leading seed accelerator company in the US. Its business model has spread to other parts of the world including Singapore.

Y Combinator is also an angel fund and puts money into selected start-ups. Since 2005 it has built up a community of over 1,600 founders and the companies in its portfolio have a valuation of over US$30 billion.[2] Its successes include AirBnB, Reddit, Homejoy, Docker, among others. Y Combinator's co-founder Paul Graham said in an interview available online: "The biggest innovation is funding start-ups synchronously instead of asynchronously, funding them all at once in a batch, instead of one at a time as they come to you. No one had ever done this before. It's better for both sides. It's much more scalable for us. They're all doing the same things at the same time. I can get up in front of them and say something once to 50 groups instead of having the same conversation one at a time with 50 groups and forgetting it half the time, right?

"But it's also better for them. It has the same advantages as mass production. Mass production, you produce things cheaper but they're also of higher quality, instead of some individual person stitching things together. They have colleagues, other people to encourage them, bounce ideas off of, compete with, exchange information about investors. It's good for them and yet, also, good for us."[3]

One of the successful accelerator companies in Singapore is The Joyful Frog Digital Incubator. Started by Hugh Mason and Meng Weng Wong,[4] it calls itself Southeast Asia's first seed accelerator, a recognition of the fact that the Singapore market in start-ups may not be large enough. Its website http://www. jfdi.asia/about describes JFDI.Asia as a community of people who practise,

---

[2] https://www.ycombinator.com. Retrieved on 7 May 2015.
[3] https://www.theinformation.com/YC-s-Paul-Graham-The-Complete-Interview. Retrieved on 7 May 2015.
[4] Cynical investors balked at accelerators. Here's how JFDI got started anyway. https://www. techinasia.com/cynical-investors-balked-accelerators-asia-jfdi-started/. Retrieved on 7 May 2015.

finance and teach innovation. "We believe that innovation need not be a mystery and entrepreneurship should not be painful or lonely. Both can be learned, working with peers and guided by mentors. Based in Singapore, we innovate primarily in Asia, for Asia. We aim to build sustainable businesses that grow value and make tomorrow better than today. Our programmes for business start-ups and professionals lead people to think and act entrepreneurially by helping them to engineer innovative businesses around their ideas. We can do this because innovation is evolving from an art into a science, and we have built a community who share their expertise and experience turning ideas into reality. Our members are entrepreneurs, investors and innovators in industry. Our partners are grassroots organisations, governments and some of the largest corporations. Our practical work is underpinned by close relationships with the world's leading academic institutions and independent thinkers." Incorporated in 2010, the innovative company has VC support from SingTel Innov8 as well as Infocomm Investments Pte Ltd, two government-backed venture funds in the IT space. JFDI is also being mentored by TechStars, a pioneer accelerator company in Boulder, Colorado, whose founder David Cohen invited JFDI to be one of the members of Techstars Network now called Global Accelerator Network.

The number of accelerators is growing. SPRING has funded four accelerators in the MedTech area from 2013. In April 2015, Infocomm Development Authority of Singapore announced that it was partnering Entrepreneur First, a British accelerator company, to help computer geeks turn their ideas into start-ups.[5] That same month, it was announced that an Australian accelerator company backed by Australian telco Telstra, Muru-D, was setting up in Singapore and looking for start-ups. The Demo Day in the accelerator model has even been turned into a TV show, a joint project between Vertex Venture Holdings and Mediacorp, Singapore's broadcast media agency. SPRING has noted that the quality of ideas and start-ups has improved in the last decade. However, the question remains whether there are so many young entrepreneurs in Singapore with so many start-up ideas that justify the growing number of accelerator companies that are also springing elsewhere in Asia. The brash, outspoken method of the Y Combinator model as well as the necessary sharing by successful entrepreneurs may not come easily to the more conservative, more secretive Asian entrepreneurs. Nevertheless, the hope is that such accelerator companies may go some way to develop an entrepreneurial culture.

---

[5] New entrepreneurship program for technology experts announced in Singapore. https://www. techinasia.com/entrepreneurship-program-technology-experts-announced-singapore/. Retrieved 9 May 2015.

Angel Investors and Venture Capitalists

A major problem with start-ups and new enterprises is running out of development funds often at a point when things are shaping up well. Traditionally,
budding entrepreneurs have turned to personal and family savings to get a
business going. In today's highly competitive, high-cost world such traditional
sources of funding are limited and limiting given the scarcity of millionaires in
most families. With the growing ubiquity of social media in the second decade
of the 21st century, crowd-funding appears to be shaping up as a way for bright
ideas to attract financial support. Millions of people worldwide contributing say,
$1 for a worthy cause can generate millions of dollars. Crowd-funding has worked
particularly well in social enterprise when like-minded people get together to
make a social enterprise work. Apart from millionaire parents, the more usual
way for a start-up to continue developing is to look for angel investors or venture
capitalists (VC). Note that angel investors are not the same as VCs with differences
that matter to start-ups. Angel investors tend to be individuals or government
agencies, are more willing to put money at the more risky early stage of start-ups
because their motivations are not always profit-driven, and amounts invested are
usually smaller. Venture capital funds, on the other hand, tend to be companies,
always profit-driven, usually not interested in risky early stage start-ups, and
can put millions of dollars into promising start-ups. Because of the large sums
involved, VCs usually want a seat on the board and a say in how the enterprise
is run. Both will have useful contacts and experience to share with the start-ups.

Venture capitalists are high-risk takers who invest in promising technology
or biomedical discoveries in the hope of striking it big. The hope of good returns
is the reason why the wealthy and governments invest in VC funds. Venture
capitalists tend to talk about successes and rarely about failures. About three-
quarters of venture-backed start-ups in the US do not return investors' capital
according to research by Harvard Business School lecturer Shikhar Ghosh.[6] It is
no wonder that venture capitalists take such a large cut when there is success.

The Singapore entrepreneurial ecosystem has a mix of both angel investors
and venture capitalist funds. Government agencies form a major part of the
angel investors while EDB has grown the venture capital industry from
$160 million in 1985 to $10 billion in 2000, and increased the number of VC
fund management companies from four to 80.[7] In 2000, NSTB and EDB

---

[6] The venture capital secret: 3 out of 4 start-ups fail. By Deborah Gage. Updated 20 Sept 2012.
http://www.wsj.com/articles/SB10000872396390443720204578004980476429190 Retrieved on
9 March 2015.
[7] Heart Work; p. 237.

released the results of their first joint survey of the VC community: by end-1999, Singapore was managing a cumulative total of $10.2 billion in VC funds. Financial services had been one of the engines of growth highlighted in the 1986 Economic Review Report. After 1985 as part of growing the financial services sector the EDB began encouraging the flow of venture capital funds to Singapore. The government also invested in such funds. The EDB started the first fund to invest in start-ups — at that time called SEED projects — in 1990. SEED Ventures was a small fund of $7.7 million and managed by Walden, one of the first three VC fund management companies to set up in Singapore, and which had picked Creative Technology as a winner very early on. VC funds in those days tended to invest only in established companies.

In 1996, NSTB set up the Technopreneurship Assistance Centre which supported start-ups in business development and marketing, intellectual property rights issues, product development and funding. In 1999, Technopreneurship21 (T21) was launched with a $1-billion Technopreneurship Investment Fund to intensify Singapore's efforts in building a conducive environment to support start-ups. The T21 framework of education, financing, regulations and facilities was designed to reinforce, promote and sustain a culture of creativity and innovation. In 2001, Technopreneurship21 was transferred to EDB, and the Technopreneurship Investment Fund to grow venture capital funds in Singapore was corporatised at the same time. Temasek Holdings, the Singapore Government's holding company for the country's equity holdings in such successful ventures as Singapore Airlines, Keppel and SembCorp, has a wholly-owned subsidiary named Vertex Venture Holdings Ltd (Vertex Group) which invests in emerging companies and leading venture capital funds throughout Greater Asia and the US. On its website, it states its mission as: "To seek out promising disruptive and transformational leaders from around the world and build them into prospective global champions with a base in Asia."[8]

The number of home-grown angel and VC funds in Singapore has increased. Besides Vertex Venture Holdings Ltd, Infocomm Investments Pte Ltd and SingTel Innov8 Pte Ltd mentioned above, NRF also catalysed a group of some 10 VC funds under its Early Stage Venture Fund programme to invest in technology start-ups, which include funds like Bioveda Capital, Extream Ventures, Walden International, Jungle Ventures, Golden Gate Ventures and Monk's Hill Ventures. Many of these VC funds also run their own accelerator and/or incubator programmes.

---

[8] http://www.vertexmgt.com/about.asp. Retrieved on 8 July 2015.

In the last decade the programmes of various agencies tasked with developing entrepreneurs take slightly different tacks with different groups of entrepreneurs. For the more established companies that want to explore new product or process development, SPRING has a fund to support capability development. It funds up to 70% of the R&D project. At the same time, SPRING links companies with the Intellectual Property Intermediary (IPI), an IP intermediation platform set up by A*STAR in 2011. IPI brings together intellectual property that are available for licensing to local companies whether from within A*STAR or from external sources. IPI is also a useful tool for start-ups that may need different bits of intellectual property to develop a better product or service. SPRING and other agencies involved in developing entrepreneurship try to create partnerships in different fields in order to catalyse different parts of the enterprise ecosystem.

One of SPRING's partners in developing an entrepreneurial culture, ACE stands for Action Community for Entrepreneurship. (See Figure 9.4) ACE was first set up in 2003 and privatised in 2014. It is a private sector movement for entrepreneurs led by entrepreneurs. ACE helps aspiring entrepreneurs start up

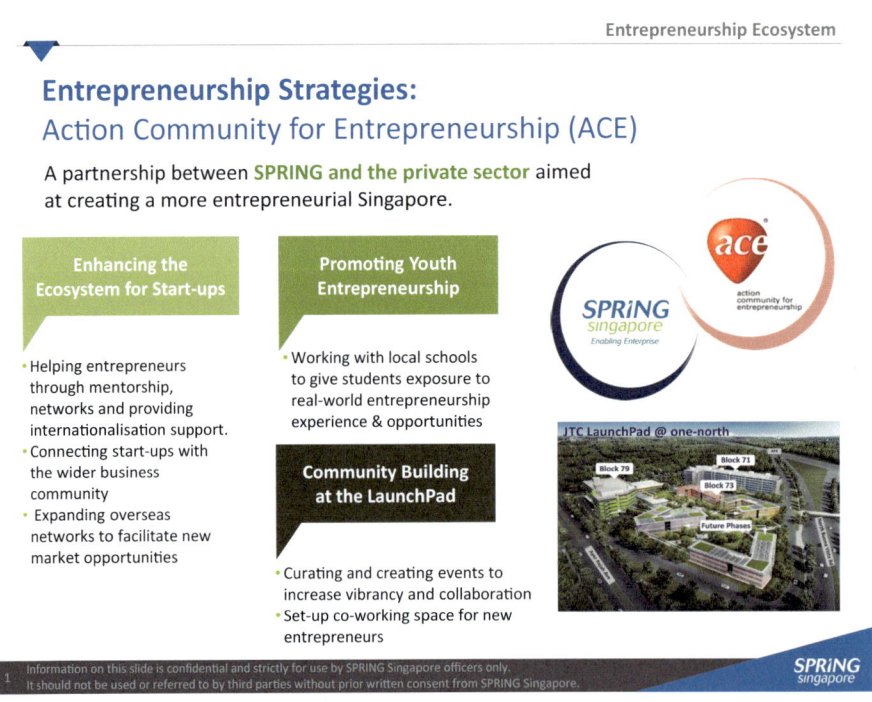

Fig. 9.4. Action community for entrepreneurship. (Reproduced with permission from SPRING Singapore).

by building a vibrant community and connecting them to resources, people and knowledge. To quote its website:[9] "ACE takes on three main thrusts to focus on strengthening the entrepreneurial scene by:

- Working with the community to engage and support promising entrepreneurs through resources, networks and mentorship;
- Being the voice to speak for entrepreneurs and lobby for relevant policy changes; and
- Reaching out to schools and youths to support their entrepreneurial efforts."

Its website says: "Located at the heart of Singapore's entrepreneurship action, JTC LaunchPad @ One-North, ACE is well-poised to be the one-stop hub to initiate a line-up of events and activities to actively engage the business community. In addition, there will be an Ideation Lab set up at ACE for aspiring entrepreneurs to get connected to facilities and valuable resources. A Visitor Centre which acts as a business concierge and gallery to facilitate site tours of this vibrant hamlet, and a web portal to showcase profiles of start-ups and partners, are also underway."

The Singapore Government has also been very supportive in developing an ecosystem for growing entrepreneurship. Speaking at the President's Science and Technology Awards dinner in 2010, Trade and Industry Minister Lim Hng Kiang said: "For 2011 to 2015, 6% of our $16.1-billion budget will be set aside to support the work of our technology transfer offices, translational and innovation centres, and enterprise incubators and accelerators. At about $1 billion, the level of support is almost a doubling from the previous budget. It will help foster a more conducive environment for our scientists to realise the commercial value of their discoveries — in the form of licenses and start-up companies, and new products and services."

Promising start-ups getting bought up by big industry players has become a possible way of further development of the companies in the technology sector, a process called "exiting". It is also one seen in the biomedical sector. Exiting is part of the economic churn. Budding entrepreneurs with bright idea start-ups do not always have to wrestle with scaling-up to become rich. For the big corporations, it is a way of buying up innovation and good ideas be it for incorporation into their own businesses or to protect their existing products from an innovative competitor. Big companies have expertise in scaling-up that start-ups still need to develop, whereas start-ups have nimbleness and flexibility that big companies rarely have. Once all the hurdles have been

---

[9] http://www.ace.sg. Retrieved 8 July 2015.

cleared and the bright idea is now finally something in production, the last stage in commercialisation is to keep the product on the market for as long as possible by staying relevant and constantly innovating.

Developing Innovation and Enterprise

Identifying and nurturing the spirit of enterprise has come to be one of the missions of the universities and polytechnics in Singapore. Tying innovation and entrepreneurship (I&E) to research is a necessary part of the end-game. Universities today, says Prof CC Hang, have a third mission in creating societal impact, and this includes innovation and enterprise (I&E). In the past, it was just teaching. Then came teaching and research. Today, universities need to embrace innovation and enterprise. It is a global trend seen in some of the best universities in the world. Academics need to understand that research plays an important role in coming up with new knowledge but also that money is being spent to create this new knowledge. Innovation is the reverse. Innovation is the question of how this knowledge is being exploited to create social goods, economic outcome. Prof Hang says: "So we are really talking about value creation now. Value comes from inventions and solution options. The mindset has to change. We need to reform our team and the team in the future so that when you really look at the important area where we want to excel, not only should we have good researchers, we should have good people who can do I&E, and they need not be the same person. So we are talking about a team. The team must have good researchers, must have good team members who can handle I&E. This is where the challenge lies. How do we educate this group of I&E people? Can they be trained? Is it something that cannot be taught in the classroom but something that you have to go out and experience? You have to go and experience failure before you actually become an expert and so on. You have to evolve. You have to learn. Every year you improve on it. It's easily another 10 to 15 years before we see results. So how do we embrace this culture of dare to dream and dare to explore, and yet be responsible? How do we get more to take risks, to take a few inventions further into innovation? Not everybody can do it, but out of 10 at least we should convince two or three to go ahead and try. This spirit of enterprise is very much like going into a new area of research. You have to invest time and effort to build a new reputation in a new area. That is the same risk as in innovation and enterprise. Academic staff have changed their research areas more than once. They have the same capability and the same spirit. Some of them should be supported to switch over to innovation and enterprise. Maybe as a university, we should start to think of importing role models. So when we invite a professor from MIT and Stanford and elsewhere, we should not just look at their publications, or look at their teaching and research because we are almost as good. We can only learn just so much more from them whereas if the same professors are very good in I&E, they will be our future role models. So we should have visiting professors of this calibre for our staff to emulate, for our students to look up to as future role models."

Promoting entrepreneurship in the universities began in the 1990s, says Teo Ming Kian in an oral history interview in 2014. Teo was chairman of NSTB from 1993 to 2001. When NSTB set up the Technopreneurship21 programme and the $1-billion Technopreneurship Investment Fund in 1999, some of the funds went into education programmes for developing entrepreneurship. One of the results at the National University of Singapore was NUS Enterprise which was established in 2001 to provide an enterprise dimension to teaching and research. It involves the university's students, staff and alumni. NUS Enterprise augments and complements the academic programmes and nurtures talents to develop an entrepreneurial and global mind-set. By igniting the entrepreneurial spirit, partnering for success and nurturing future entrepreneurial leaders, NUS Enterprise aims to make a positive impact on Singapore's economy and beyond. This is done through three key thrusts: Entrepreneurship Support, Experiential Education, and Industry Engagement and Partnerships. Its website says: "We provide experiential entrepreneurial education, active industry partnerships, comprehensive entrepreneurship support, catalytic entrepreneurship outreach, and being Asia's thought leader in innovation and enterprise, as well as the bridge to industry for our university." In 2011, it went into a joint enterprise with other government agencies to set up Block 71.

The A*STAR RIs have been established as mission-oriented knowledge entities to attract, train and retain RSEs and to deepen the research capability of the nation. They have played a significant role as innovation partners of MNCs, large local enterprises and SMEs. Some of their enterprising RSEs have contributed more directly in innovation and entrepreneurship by creating successful spin-off companies. A*STAR has created a portfolio of 54 start-ups between 2011 and 2014. That's more than 13 a year or an average of more than one a month! In the physical sciences and engineering field, an example is Singular ID, which spun off in 2005 from the Institute of Materials Research and Engineering to commercialise A*STAR's nano-magnetic fingerprint technology to combat counterfeiting. Initially operating in the automotive and luxury goods industries, Singular ID was acquired in 2007 for $20 million by an MNC, Bilcare Ltd, to support its pharmaceutical packaging solutions and ensure the authenticity of pharmaceutical products. Singular ID was renamed as Bilcare Technologies.

Another notable physical sciences and engineering example is Zweec Analytics, a spin-off from A*STAR's Institute for Infocomm Research (I²R) that is working on the kind of high-tech advancements that make a real difference to Singapore. Based on I²R's media and behavioural analytics expertise, Zweec's Fish Activity Monitoring System uses groups of freshwater fish to detect changes in water quality. The technology functions on the basis that changes in water quality affect the motion of fish, which will prompt the system to send an alert to a central monitoring location. This technology acts as a first line of defence against contaminants in water, allowing faster response to changes in water quality. While the use of fish in water-quality management is not new, Zweec improved on existing technology, creating a reliable algorithm-based system that is fully automated. Not only has the system been deployed in Singapore by PUB, it is also being used around the region for water quality assurance, notably in Taiwan and China.

In the Biomedical Sciences field, one representative example is MerLion Pharmaceuticals Pte Ltd. It was formed in 2002 through the privatisation of the former Centre for Natural Products, a unique unit of IMCB. It has recently focused on developing its lead antibacterial candidate, finafloxacin. In February 2015, this candidate won FDA approval in the US. It demonstrates that Singapore can develop a drug that wins difficult US FDA approval. And there are two more candidates awaiting approval, plus a healthy pipeline of candidates.

A*STAR has also seen many successes in the sphere of diagnostics. One example to highlight is PathGEN Dx, a spin-off in 2011 from the Genome Institute of Singapore, which has developed the PathGEN® PathChip test-kit that can detect the presence of more than 70,000 viruses and bacteria from human samples. Another is Veredus Laboratories, which with A*STAR's Singapore Immunology Network, developed the VereTrop™ lab-on-chip diagnostic kit that can identify 13 different major tropical diseases — including dengue fever, malaria, chikungunya, and hand, foot and mouth disease — from a single blood sample. Amid international concern about the spread of the Middle East Respiratory Syndrome (MERS-CoV), the diagnostic capabilities of PathGEN Dx and Veredus to detect the MERS virus are also being tapped on by private laboratories and research institutions.

Intellectual Property as Assets

As part of the knowledge-based economy, it must not be forgotten that intellectual property can be looked on as economic assets that are tradeable. Hence, one could say that public research institutes are producing economic value albeit of an unusual kind. While GET-Up and T-Up were being developed to help SMEs, in 2002 Exploit Technologies Pte Ltd (ETPL) was formed as a wholly-owned subsidiary of A*STAR, to help bridge the gap from mind to market. Speaking to *The Straits Times* at its launch, then A*STAR chairman Philip Yeo said: "It's tax payers' money. We are not Father Christmas. We're very economically orientated. We are not doing science for science's sake. So we have to try to commercialise the ideas."

The founding executive chairman of ETPL until 2009 and concurrently A*STAR's managing director until 2006 was Boon Swan Foo. His task was to identify, protect and exploit promising intellectual property created by A*STAR's research institutes. He said in the 2006/7 A*STAR annual report: "The philosophy of patenting and knowing how to claim their due right in research is something we have inculcated in A*STAR. We help scientists patent and publish quickly. As of today, ETPL manages a portfolio of more than 3,400 active patents and applications.

Handling the IP of all the A*STAR research institutes, ETPL works closely with SPRING, International Enterprise Singapore, Infocomm Development Authority, Media Development Authority Singapore and Workforce Development Authority. Its website, https://www.etpl.sg/innovation-offerings/technologies-for-license, sets out some of the technologies available for licensing in an interesting array of fields that range from food to materials to medical devices to chemicals to electronics. Between 2011 and 2014, ETPL signed 825 licenses for A*STAR technologies, with an expected commercial value of $595 million. Its successes include the SARS detection kit, slip-resistant glass, and putting DNA on a silicon chip. Through ETPL's licensing of IP to industry partners, SMEs get critical boosts to their business while achieving commercial impact for the RIs and realising value capture for A*STAR's work.

Boon said in A*STAR's 2011 commemorative publication: "Invention is just one part of the innovation chain. Obviously the better the invention or the more significant the breakthrough, the higher the chances of success, and the more markets and alternatives you have. But you must also constantly innovate in terms of use (by coming up with new products), markets (by finding new growth areas), and product development (by making your product better than your competitors). Sometimes the exact placement of a feature is what decides if your product is a winner or not."

A key mechanism that ETPL employs to bridge the gap between lab and market is its gap funding scheme which ETPL uses to raise the technology readiness level of commercially promising A*STAR research to a level closer to market needs. With the help of this funding, A*STAR's research institutes can carry out technology development and refinement based on their inventions and proof-of-concept prototypes over a period of up to 12 months. The scheme aims to create a pool of market-ready technologies for commercial applications, refinements for mass production and enhancements. Since the launch of Gap funding, over 100 gap projects worth over $70 million have been completed. These have been commercialised at a rate of 2.8 licenses per project, with an expected commercial value in excess of $410 million. The completed projects have also resulted in 44 start-ups which in turn have raised more than $75 million in follow-on funding.

SMEs need to be savvy about the value of intellectual property as critical to entrepreneurial survival as the story of Trek 2000 and the ThumbDrive illustrates. (Box: Trek 2000 and the Importance of Patent Protection, page 192) SMEs need to be made more aware of the potential of IP as a competitive strategy to go beyond just protecting themselves but to also use IP to grow beyond Singapore, to grow in the region and globally. The universities' engineering and business schools have introduced IP case studies with the help of the IP Academy to prepare students for working with SMEs. Educating future SME leaders and managers on the importance and value of IP is a way to spread knowledge about the value of IP. IPOS also has a one-stop IP customer service centre that provides basic advisory on IP filing in Singapore and overseas, among other things. It organises seminars on common IP matters, legal strategies and real case studies. It has launched the "IP Business Clinic" to provide professional advice on IP business strategies including IP monetisation, commercialisation and protection of IP overseas.

While educating SMEs on the value of IP is an ongoing project, another project is being undertaken to position Singapore as a regional IP hub, a project being undertaken by EDB, Ministry of Trade and Industry, and Ministry of Law. IP has become an economic development activity and has led to the development of new professional expertise that revolves round patenting, examination of patents, protection of IP, and education on IP. (See Chapter 3) As a small country, it is not possible for Singapore to develop a wide spectrum of expertise in patent examination. Thus, it has chosen to specialise in one or two areas that pertain to a key manufacturing sector and research interest. Said Prof CC Hang, founding chairman of IPOS: "We are not like UK or Japan but we should choose one or two areas, each area maybe with 50 to 100 experts, and say that, yes in these two areas we will do our patent examination. And with that, we are going to create these expert examiners. And because we are getting our feet wet in examining other people's patents, patents in the region will come to us. Then we can have one more bullet in strengthening ourselves as an IP Hub. At least in certain areas, we can claim that we are the regional centre." It is a major area of development that is still in the works but one that can be built on Singapore's growing strength in S&T and established reputation for probity.

Think Local, Act Global

For Singapore to play a role in innovation and enterprise creation in the 21st century, it has to go beyond technology which is only a starting point. It must have access to IP protection, IP exploitation. These are the minimum conditions that a country must have but they are not sufficient for innovation and enterprise creation. In today's world it is about design-driven innovation, first-rate business models, opportunity identification, spotting the needs of the market place. The business school mantra "Think global, act local" has applications in R&D in the globalisation era of the past 30 years. In future, it will be supplemented by a new paradigm of "Think Local, Act Global". There are technological solutions that may be too basic for advanced economies but have relevance in developing economies or what is sometimes called good-enough technology. This is an economic opportunity for Singapore enterprise to couple its geographical position in a developing region with a world-class R&D capability. Entrepreneurs need to look for upstream technology that they can exploit for downstream applications. This is how to think local (to identify real high-growth market needs) and act global (to source for affordable, good enough technologies from advanced countries with appropriate R&D). And such solutions may in fact be exported back to the advanced economies as "reverse innovation" because it is often more cost-effective.

The question that comes up is whether there is sufficient understanding or knowledge about local conditions outside Singapore for budding Singapore entrepreneurs to identify such opportunities. There is a need to build up this knowledge. One of the ways that Prof Hang proposes is to send students to study in the regional universities, to build local ties and local knowledge. He is of the opinion that there is now more to be gained from going to regional universities than from US and UK universities now that the knowledge gain from attending universities in these advanced countries is flattening. The number of student exchange programmes between Singapore and regional universities needs to be increased. In 2012, EDB sponsored the setting-up of the Institute of Asian Consumer Insights (ACI) at Nanyang Technological University. The press release announcing the launch stated: "The Institute, funded jointly by the Singapore

Economic Development Board and Nanyang Technological University (NTU), is a world-class, first-of-its-kind institute focused on Asian consumers. Through research, education programmes and industry collaborations, ACI will help companies develop strategies for Asian markets. Hosted at NTU and funded up to $77 million over five years, the Institute will help businesses innovate brands, products and services, based on insight about pan-Asian consumer needs, wants, and preferences." At the top of its list of to-dos is an Asia Consumer Summit organised with the *Financial Times* and modelled after the World Economic Forum in Davos. This and other conferences draws together global business leaders and is a platform for learning more about Asian consumers.

More can and should be done at a lower level to build up local insights into the region and the rest of Asia in cohorts of young Singaporeans, says Prof Hang who thinks that such networks and ties will serve Singapore well in the 21st century, just as such networks and ties with the more developed world have served the country well in the 20th century. Not only would this add to their value proposition for their future MNC employers but also open up opportunities for economic enterprise.

The Years Ahead

R&D will continue to be a key strategy for growing the economy and to ensure its competitiveness. It helps create solutions to improve people's lives, as well as job opportunities and new industries. It will continue to make Singapore relevant to its partners be they corporate or national partners. Singapore's R&D landscape has come a long way since the 1980s when R&D was identified as the way to restructure the economy. It has been amply shown that Research, Innovation and Enterprise are contributing not just economic impact but changing the way Singapore's leaders see the goals of education, the generation of creativity and the numerous ways in which S&T can improve and change lives. Indeed even in the short space of just 25 years as policies, programmes and initiatives pegged to R&D and S&T have multiplied, the buzz and vibrancy in Singapore's research ecosystem has put Singapore on the global R&D map. In the years ahead, staying on the map calls for more and even better science and engineering, and upholding the country's reputation for high standards in R&D. (See Figure 9.5)

NRF sees R&D as very important in providing Singapore with solution options beyond economic impact. The Government's role is to harness innovation to provide solutions to fundamental challenges that Singapore must address to win the future. Its role is not to place bets on any particular industry;

National Innovation Challenges

Energy National Innovation Challenge
- To develop cost-competitive energy solutions for deployment within 20 years to help Singapore improve energy efficiency, reduce carbon emissions and increase energy options

Land & Liveability National Innovation Challenge
- To create new space cost-effectively and optimise the use of space to sustain Singapore's long-term growth and resilience

NATIONAL RESEARCH FOUNDATION
PRIME MINISTER'S OFFICE SINGAPORE

Fig. 9.5. Addressing future challenges in Singapore. (Reproduced with permission from National Research Foundation).

but rather to answer specific questions that are vital to the country's future, and in answering those questions, provide solutions to the world. For example, if sea water levels rise because of global warming what options does an island-nation like Singapore have? Or if the nation needs to turn to nuclear technology, what is the technology that it should adopt to ensure the necessary levels of safety? If Singapore needs to expand its limited land size and move underground to create more living spaces, what are the options? When the decision was taken to develop the Jurong Caverns specifically as underground facility storage for ammunitions, it took some 15 years to be realised. Before such major decisions there are a lot of studies and exploration of the options. Thus, when a decision is made finally, it is an informed decision. The solution options must be ready when they are needed as R&D takes time. Research and innovation can seed the solutions for doing well in a more resource constrained, more crowded, more economically uncertain planet. It can re-imagine fundamentals, the foundations of how all economies function, the way we feed ourselves, the way we transport ourselves, the way we get water, the way we power our lives, that way we create economic value. By focusing on delivering solutions for Singapore — how do we grow and maintain a good life with fewer resources — we can offer answers to the world. The role of R&D will continue to be very important to Singapore and what is of concern is whether we will continue to attract the brightest and youngest people to be engaged with it. Ultimately, success rests on people and talent. The backbone of a technological, knowledge-driven economy is a healthy scientific community. Sustained growth and stability of our R&D talent base are essential to our national research enterprise. It is up to our next generation of scientists and technologists to build on the foundations that have been laid, to scale new heights in the next phase of Singapore's R&D journey. They will be working on big important problems, and shaping the future of Singapore — and in some way contributing to the global availability of options in an ever more constrained world totally dependent on science and technology for growth and even survival.

Appendices

Singapore's Science & Technology, R&D Timeline

1961: Setting up of Economic Development Board (EDB) with Hon Sui Sen as chairman to draw investors/industrialists to Singapore; start of the construction of Jurong Industrial Estate.

1963: Setting up of Industrial Research Unit within EDB with the help of Colombo Plan aid from New Zealand.

9 August 1965: Singapore separated from Federation of Malaysia to become an independent republic. Singapore's per capita income was US$800.

1965: Engineering diploma course at Singapore Polytechnic became a degree course.

1966: Industrial Research Unit became a full member of International Organisation for Standardisation; the first Singapore Standard issued was on timber and primers.

1967: Science Council formed to push for national advancement of S&T and build up human resources in scientific R&D.

1967: Singapore became a member of the Commonwealth Scientific Committee whose main goal was to bring about cooperation between Commonwealth countries in science and technology.

1967: Singapore became a centre for the International Agency for Research of Cancer.

1967: UNESCO team led by Dr Kurt Billig and D Merrill in Singapore to assess the need and potential for technological research facilities.

1968: Creation of the Ministry of Science and Technology with Dr Toh Chin Chye as the first Minister tasked with restructuring technological education.

1968: Science Council organised the landmark National Conference on Scientific and Technical Cooperation between Industries and Government Bodies to identify areas of mutual interests and to consider ways and means to foster cooperation. Involving more than 200 local scientists, technologists, industrialists and economic planners, this conference set the agenda for the upgrading of industry through science and technology, technical manpower development, and the direction of R&D in Singapore.

1968: Setting up of National Industrial Training Council to make policy for technical education and industrial training to meet manpower requirements of new industries.

1969: Major re-organisation of science and technology agencies as well as science and technology education by the Ministry of Science and Technology. Establishment of the School of Postgraduate Medical Studies with the first Masters of Medicine degrees in Surgery, Paediatrics, and Internal Medicine offered in 1970.

1969: Science Council proposed the creation of Science Centre.

1969: Placement Agency set up by Science Council to match graduates in science and engineering with industries as well as create a database of overseas undergraduates expected to return to Singapore. The Agency also began working with EDB and Economic Research Unit to set up machinery to match the supply and demand for technical manpower at the mid- and higher levels.

1969: International Atomic Energy Agency Committee awarded University of Singapore a research contract worth US$3,000 for its proposal to study "productivity and bioenergetics of tropical ponds (including fish ponds)" in Singapore, the first research contract awarded to Singapore.

1969: Standards Council formed; Industrial Research Unit of EDB renamed Singapore Institute of Standards and Industrial Research (SISIR), with Dr Lee Kum Tatt as its head.

1970: Science Council submitted *A Preliminary Report on R&D Activities in Singapore* to the Ministry of Science and Technology. The report was prepared by Dr Meir Ben Zvi of Israel's National Council for Research and Development.

1971: Science Council instituted the Science Teachers' Awards to recognise inspiring science teachers. Applied Research Fellowship Scheme introduced to foster links between scientists and engineers in industry with their counterparts in educational institutions. Fellowships were awarded for specific research projects which could be part-time in nature. From 1975, these research projects had to be full-time research projects. The scheme was phased out in 1977.

1971: SISIR introduced the "Made in Singapore" mark for products that had met its standards. The SISIR Certification Mark Scheme was actively promoted to get local industries to raise the quality of their products. By 1983, 700 Singapore-made products carried the SISIR mark. SISIR also initiated the Singapore Quality Reliability Association to promote quality and reliability awareness amongst manufacturers.

1971: Ministry of Science and Technology published the first *Directory of Scientific and Technical Research and Consultancy Establishments in Singapore*. Besides SISIR, research establishments comprised mostly engineering consultancies.

1971: Ministry of Science and Technology conducted the first triennial National Survey of Scientific Manpower.

1971: University of Singapore's Faculty of Engineering and Institute of Engineers conducted the first survey of engineering manpower.

1972: Dr Goh Keng Swee, Minister for Defence, set up Project Magpie also called Electronics Test Centre with three hand-picked young engineers to form a defence technology R&D group that would later develop into Defence Science Organisation, Singapore's first R&D body.

1972: Public Utilities Board came up with its first Water Master Plan.

1972: Science Council organised the first Science Quiz 1972 in collaboration with Radio & Television Singapore, Ministry of Education, and Science Teachers' Association with Dr Tay Eng Soon as the quiz master.

1972: Singapore became a World Health Organisation Collaborating Centre for Upper Respiratory Tract Tumours. Other specialities were subsequently added: Enteroviruses, Immunology, Maternal and Child Health/Family Planning, Health Informatics.

1972: Beechams Pharmaceuticals set up a plant to produce semi-synthetic penicillin from imported penicillin nucleus. In 1976, Beecham put in equipment to produce the penicillin nucleus itself, the first in the world to use the latest technology for the manufacture of the penicillin nucleus.

1973: National Productivity Board set up with Augustine HH Tan appointed as chairman.

1973: Industrial Training Board set up with goal of coordinating and beefing up industrial training and development of technical manpower at craftsman and skilled worker level.

1973: Setting up of Applied Research Corporation as a non-profit company and institution of public character to offer small and medium-sized enterprises (SMEs) a wide range of services in economic science, business management, engineering systems and industrial R&D.

1973: SISIR became an autonomous statutory board, with Dr Lee Kum Tatt as the inaugural chairman.

1973: SISIR set up an Instrumentation Services Centre initially focused to develop, verify, calibrate, repair and service instruments and equipment and certify that the instrument or equipment met the standards set for it. In 1976, it set up the Temperature Standards Laboratory to maintain temperature standards; in 1977 it built an engineering metrology laboratory to support the growing precision engineering industry with plans to upgrade as the industry sector grew; in 1979 it embarked on an upgrading programme. In 1987, when it moved to purpose-built metrology labs in Science Park, it came to be called the Metrology Centre by the media. The word "National" was subsequently added to its name. In 2008, it became an A*STAR centre.

1974: Ministry for Science and Technology conducted the first triennial National Survey of Engineering Manpower, and the second triennial National Survey of Scientific Manpower.

1975: Science Council and Radio and Television Singapore collaborated to produce a series of science documentaries titled *How and Why*, another of the Council's public education programmes, one of Singapore's first home-grown science documentaries.

1976: Ministry of Science and Technology conducted a survey on public and private sector of establishments that did any kind of R&D. Some 250 establishments took part and just over 16% or about 40 establishments were found to be engaged in some form of R&D activity. Some 541 projects were carried out with total expenditure of $5.3 million, the bulk of it by public sector. Measured against 1975 GDP of $13,681 million, total expenditure on R&D was only 0.04% of GDP.

1976: SISIR incorporated Setsco Pte Ltd to commercialise the Institute's research results and to provide other technical and consultancy services to industries. Its first commercial product was the RISIS Orchid. Setsco Services Pte Ltd was incorporated in 1981, also with similar objectives.

1977: Opening of Singapore Science Centre as one of the ways of exposing young people to the marvels of science and technology. Science Council moved its offices to Science Centre.

1978: First National Survey on R&D conducted by Science Council.

1978: EDB introduced the Product Development Assistance Scheme with financial grants to local SMEs to improve manufacturing processes or for developing new products. The dollar-for-dollar grants could be used for engaging R&D assistance from agencies such as SISIR or Applied Research Corporation.

1979: Development Division of Ministry of Finance became Ministry of Trade and Industry with its original function of advising Government on the overall economic and manpower strategy remaining unchanged.

1979: SISIR's metrological services were upgraded to meet the higher-technology needs of the electronics, aerospace and precision engineering industries.

1979: Establishment of Skills Development Fund to encourage employers to upgrade the skills of their staff, following a recommendation of the National Wages Council.

1980: Incorporation of Everbloom Mushrooms, the first spin-off from academic research started by NUS biochemist Dr Tan Kok Kheng with funding from family members and EDB's Small Industries Finance Scheme, with the company going into full production in 1983.

1980: Merger of University of Singapore and Nanyang University to form National University of Singapore (NUS) and the setting up of Nanyang Technological Institute (NTI) on Nanyang University's campus in Jurong. Dr Tony Tan, Senior Minister of State for Education, became its first Vice-Chancellor. As part of NUS Faculty of Engineering, NTI started with three schools of engineering: Civil and Structural, Electrical and Electronic, and Mechanical and Production. In 1991, NTI was given a university charter and became Nanyang Technological University (NTU).

1980: NUS had 171 postgraduate research students as part of plan to increase postgraduate numbers. In 1995, postgraduate research students numbered 1,500 with majority in faculties with strong research emphasis, namely Engineering, Medicine and Science.

1981: Formation of the National Computer Board chaired by Philip Yeo and the launch of the national computerisation programme on the recommendations of National Committee for Computerisation.

1981: Institute of Systems Science formed primarily as a teaching institute. Its R&D programme started in 1986.

1981: Apple Computer became the first company to manufacture PCs in Singapore for world markets starting with printed circuit board assembly. Disk drive makers wooed by EDB.

1981: Introduction of Research and Development Assistance Scheme (RDAS) to be administered by Science Council, a grant scheme for R&D projects of national significance.

1981: Ministry of Science and Technology closed and its functions transferred to other ministries. Ministry of Trade and Industry took over its R&D promotion functions and boards such as Metrication Board, Weights and Measures, and the technical manpower surveys; Science Council and technical education went to Education; and Department of Scientific Services returned to Health.

1981: Sim Wong Hoo and Ng Kai Wai set up Creative Technology as a computer repair shop, releasing its self-designed computer, Cubic 99, in 1984 and Cubic NT featuring color graphics and stereo sound reproduction in 1986.

1982: Seagate Technology, inventor of the hard disk drive, became first disk drive manufacturer to set up manufacturing facilities in Singapore. In 1984, it set up a facility at Science Park to do development work.

1982: Glaxo set up a plant to manufacture the active compound, Rantidine Hydrochloride, for its new drug, Zantac, discovered in the 1970s. The Singapore Glaxo plant was the first plant here to get a US Food and Drug Administration inspection for which it got a clean bill of health.

1982: Department of Obstetrics and Gynaecology of the Faculty of Medicine of NUS did Asia's first successful in-vitro fertilisation and embryo transfer. The baby, a boy, was born on 19 May 1983.

1982: SISIR set up the Materials Technology and Application Centre to help industry make better use of raw materials and assist local industries that were part of the supply chain of high-tech MNCs.

1982: Science Council mooted plans for Science Park in Kent Ridge and its subsequent development handled by EDB and JTC.

1982: Science Council set up a study group to look into the upgrading of Singapore's patent system, inviting a four-member World Intellectual Property Organisation Technical Mission to Singapore for consultations.

1983: Between 1983 and 1987, Science Park took in 34 R&D organisations.

1983: Science Council organised an ASEAN-EEC Seminar on *Biotechnology — The Challenges Ahead*, in which the keynote speaker was Dr Sydney Brenner, Director of the UK Medical Research Council.

1984: Singapore Biotech Pte Ltd to handle a wide range of scientific and medical products incorporated in 1983 with Temasek Holdings Pte Ltd, US Summit Corporations, DBS Bank Ltd and Intraco Ltd as shareholders. Its research division (Singapore Biotech Research Laboratories) would focus on diseases prevalent in Southeast Asia.

1984: SGS Singapore, a subsidiary of Italian semiconductor manufacturer SGS Microelecttronica, opened the first semiconductor wafer diffusion plant in Ang Mo Kio Industrial Park 2. The man who made this decision, Pasquale Pistorio, became one of the first two honorary Singapore citizens when the honour was introduced in 2003.

1984: Petrochemical Corporation of Singapore (Sumitomo Chemical) and its first downstream affiliates — Philips Petroleum Singapore Chemicals, Polyolefin Company (Singapore) and Denka Singapore plants started up on Pulau Ayer Merbau.

1984: Official opening of Science Park with Norwegian company Den Norske Veritas Marine Technology Centre as the first tenant.

1985: Diagnostic Biotechnology Pte Ltd established in Science Park to research, make and distribute diagnostic products worldwide, focusing on immunodiagnostics. Its product range included enzyme immunoassays (ELISAs) and Western Blots for the AIDS viruses, HIV1, HIV2 and HTLV1, and for the dengue fever virus.

1985: Science Council set up two panels of international advisors, one on physical sciences and engineering, the other on biological sciences, the first-ever international panels of advisors.

1985: Grumman International NTI CAD/CAM Centre (GINTIC) set up with Dr James Boyd as its inaugural director. Later called GINTIC Institute of Manufacturing Technology which merged with Institute of Manufacturing Technology in 1993 to become Singapore Institute of Manufacturing Technology (SIMTech).

1985: Setting up of Institute of Systems Science (ISS) in NUS.

1986: Information Technology Institute set up as applied R&D unit in the National Computer Board, as part of the national computerisation programme.

1986: Publication of Report of the Economic Committee titled *The Singapore Economy: New Directions* to look at future economic strategy in the light of 1985 downturn.

1987: EDB formed National Biotechnology Committee to look into the development of biotechnology as economic strategy.

1987: Institute of Molecular and Cell Biology with Dr Christopher YH Tan as its inaugural executive director and Dr Sydney Brenner as its advisor, became operational.

1987: Opening of Pacific Biomedical, Singapore's first artificial heart valve maker.

1987: Science Council inaugurated Technology Month to showcase S&T.

1987: SISIR organised its metrological services into the National Metrology Centre.

1987: Singapore Institute of Standards and Industrial Research moved to its $60-million building at Science Park, and at the same time announced the setting up of an R&D Incubator Centre in the building. The press began calling the metrology laboratories here the Metrology Centre.

1987: NUS joined BITNET, the computer network that was connecting universities and R&D establishments in the US and Europe, marking Singapore's entry as a fully fledged member of the international networking community, a move which led directly to its present status today as a major Asian Internet node.

1988: Announcement of National Biotechnology Plan to develop biotechnology and release of National Biotechnology Master Plan in 1989 which mapped out five strategies to enhance level of biotechnology in Singapore: technology, manpower, industry and infrastructure development, and in promotion of public awareness and interest in biotechnology.

1988: World's first successful micro-injection pregnancy carried out by the Department of Obstetrics and Gynaecology of NUS Faculty of Medicine. The technique called Micro-Semination Sperm Transfer enable men with low sperm count to father their own children. In 1992, the same department chalked up another first by achieving a human pregnancy from the direct injection of the sperm into the cytoplasm of the egg.

1988: NUS set up Surface Science Laboratory for multidisciplinary research work, with industry collaboration its major concern.

1988: NUS transferred its Microprocessor Applications Centre to SISIR in a move to consolidate SISIR's capabilities in engineering product design and development. At SISIR, a new Electronics Test Centre was also established to provide highly-sought testing and evaluation services to the electronics industry and the growing number of International Purchasing Offices in Singapore.

1988: Sim Wong Hoo set up Creative Technology subsidiary, Creative Labs, in Silicon Valley, United States, and began marketing the Game Blaster sound card.

1989: Landmark meeting chaired by Minister for Education Dr Tony Tan with representatives from the universities, EDB, Ministry of Trade and Industry to discuss setting up of advanced engineering research institutes, beefing up of R&D and the eventual setting up of the National Science and Technology Board.

1989: Life Technologies commercialised the world's first AIDS automatic diagnostic system which was developed in SISIR's R&D Incubator Centre.

1989: Creative Technology released the Sound Blaster sound card, revolutionising the PC and making possible its evolution into an entertainment unit. In 2006, Creative Technology sued Apple over its use of Creative's patented interface technology in Apple's iPod players. The case was settled out of court in Creative's favour to the tune of US$100 million.

1989: Release of National Automation Plan to tackle labour shortage. Robot density rose from 9.7 in 1987 to 40.8 in 1990, making Singapore third after Japan and Sweden in industrial use of robots.

1990: Formation of Bioprocessing Unit, renamed Bioprocessing Technology Centre in 1995, and in 2003, becoming Bioprocessing Technology Institute.

1990: NUS established the Postgraduate School of Engineering to spearhead expansion of the Engineering Faculty's postgraduate programmes.

1991: Establishment of National Science and Technology Board (NSTB) and announcement of 1st National Technology Plan. Also announced was the establishing of a computer network called TechNet being developed by NUS to link Singapore's R&D community with one another and counterparts worldwide by 1992. TechNet was part of the $2-billion National Technology Plan.

1991: SISIR convened the inaugural meeting of its International Panel of Advisors on Technology to help the Institute map out its Technology Master Plan.

1991: Nanyang Technological Institute became Nanyang Technological University.

1991: Opening of SISIR's Food Biotechnology Centre.

1991: SISIR elected to the governing council of the International Organisation for Standardisation at its General Assembly in Madrid, Spain, for a three-year term starting in 1992.

1991: SISIR introduced its Soft Start R&D programme to help companies make trial starts in R&D without having to make heavy upfront investments.

1991: Establishment of Institute of Microelectronics.

1992: Setting up of Magnetics Technology Centre, later to become Data Storage Institute in 1996.

1992: Creative Technology became the first Singapore technology company to be quoted on NASDAQ (National Association of Securities Dealers Automated Quotations).

1992: NUS Medical Faculty initiated the Advanced Specialty Training programme for post-MMed doctors. In 1995, more than 230 doctors from 15 countries were pursuing the MMed degree in 11 disciplines, with another 100 doctors doing the Advanced Specialty Training Programme.

1992: Motorola set up its Asian Motorola Manufacturing Systems facility to carry out manufacturing R&D.

1993: Launch of Singapore's regionalisation drive: establishment of ASEAN Specialised Meteorological Centre in Singapore.

1993: SISIR launched Thin Film and Surface Technology programme to position Singapore as regional R&D centre for advanced materials. The programme focused on precision optics coatings, critical to the development of new coating materials including the next generation of optical thin films.

1994: SingTel set up Singnet, Singapore's first Internet Access Service Provider (IASP); Sembawang Media, ST Computer Systems and Singapore International Media formed Pacific Internet Consortium to buy up Technet and commercialised its services as Pacific Internet Corporation Pte Ltd in 1995.

1994: Setting up of the Process Analysis and Optimisation Enterprise (PAOE) in NUS that would in 1998 be upgraded into the Chemical Process and Engineering Centre, and in 2000 the Institute of Chemical and Engineering Sciences.

1994: Philips set up its R&D Centre for Manufacturing Technology, the first outside of the Philips HQ R&D facility in Eindhoven with an active R&D component.

1994: CarnaudMetalbox set up regional R&D centre for packaging technologies with a $32-million R&D programme.

1994: Shell announced setting up of Shell Research Eastern Pte Ltd to embark on $25-million R&D in areas of fuels, lubricants and bitumen.

1994: ASM Technology Singapore boosted its technological capabilities in the design of semiconductor assembly and packaging equipment with a $60-million R&D programme.

1994: Setting up of National Medical Research Council to oversee the development and advancement of medical research in Singapore.

1995: SGS-Thomson Microelectronics Asia Pacific announced three strategic technology research programmes in DSP, wafer processing and advanced IC packaging technology costing some $42 million.

1995: SISIR set up the National Patent Information Centre as a one-stop patent information resource centre to support R&D in Singapore. In addition, it was to promote activities to educate industry and general public on the use of the patenting system.

1995: Patents Act came into force allowing the primary filing of patents in Singapore.

1996: Productivity and Standards Board formed by merging Productivity Board and SISIR.

1996: Centre for Signal Processing opened at Nanyang Technological University to strengthen Singapore R&D in digital signal processing.

1996: GEA-NUS Pharmaceutical Processing Research Laboratory set up to develop capabilities for the secondary manufacturing of pharmaceuticals. This laboratory offered industry a research facility with pilot scale equipment to improve their processes and products. Traditional medicine companies like Eu Yan Sang International, Tithes Marketing, and Healthaid Enterprise used the laboratory to modernise their products and bring them closer to the needs and preferences of the new Asians.

1996: Motorola set up advanced R&D facilities for paging and display technology.

1996: Thomson Multimedia set up R&D facilities to develop digital products.

1996: Panasonic launched corporate laboratory here.

1996: Lucent Technologies set up new regional telecommunications R&D facilities.

1996: Flexera set up Singapore's first chip-scale packaging materials development centre.

1996: BP opened $20-million Asia-Pacific Regional Technology Centre to conduct R&D into specialty oils.

1997: Setting up of Institute of Materials Research and Engineering.

1997: Defence Science Organisation incorporated as a not-for-profit company called DSO National Laboratories.

1997: Lilly-NUS Centre for Clinical Pharmacology set up.

1998: Hewlett-Packard announced investments of US$100 million over next five years for R&D on office inkjet printers.

1998: Glaxo-Wellcome opened a new product manufacturing and development facility.

1998: NUS Department of Pharmacy opened GEA-NUS Phramaceutical Processing Research Laboratory jointly with GEA/Nio Group from the UK to undertake research and provide manpower training for local industry.

1998: Local company Hydrochem announced further commitment to research in water treatment processes with the building of a pilot plant to be tested in different industries over two years.

1999: A Genetic Modification Advisory Committee formed as a joint effort involving research bodies, EDB, Primary Production Department, health and environment ministries and hospitals to address public concerns over genetically modified food.

1999: Local company Biosensors International which makes state-of-the-art surgical instruments increasing its R&D spending to make its products safer and more practical for heart surgery.

1999: Compaq CustomsSystems opened its new Engineering Design Centre to build specially engineered hardware systems for worldwide customers.

1999: Giesecke & Devrient, a German company doing printing of banknote and securities, set up its first regional R&D centre to develop core capabilities in smart card technologies for the region.

2000: Singapore 2000 Global Technopolis provided backdrop for second Global Strategies Conference with its focus on world-class partnerships.

2000: Ministry of Defence established the Defence Science and Technology Agency as a statutory board to provide a more nimble operating framework for exploiting science and technology for defence and the mission of the Singapore Armed Forces.

2000: Singapore Genome Programme organised, becoming the Singapore Genome Institute in 2001.

2000: Formation of Bioinformatics Institute as an IT services and support unit, becoming a biological research institute in 2007.

2002: National Science and Technology Board renamed Agency for Science, Technology and Research or better known as A*STAR.

2002: Trade Development Board renamed International Enterprise Singapore; Productivity and Standards Board renamed Standards, Productivity and Innovation Board (SPRING Singapore).

2002: Trek 2000 International granted Singapore patent for the USB ThumbDrive, with the patent extending to 30 other countries.

2003: A*STAR established the Growing Enterprise with Technology Upgrading (GET-Up) programme to transfer technologies and researchers to SMEs.

2003: Setting up of the Institute of Bioengineering and Nanotechnology, the world's first.

2003: Introduction of Honorary Citizenship for people who have made outstanding contributions to the development of Singapore. Pasquale Pistorio of SGS Miroelecttronica and Dr Sydney Brenner dubbed the "Father of Singapore Biotechnology" became the first two honorary citizens of Singapore at the same ceremony.

2004: Formation of Centre for Molecular Medicine, becoming Institute of Medical Biology in 2007.

2005: Singapore Stem Cell Consortium set up.

2005: Singapore Bioimaging Consortium set up.

2005: Duke University and NUS signed an MOU in which the two institutions would partner to establish the Duke-NUS Graduate Medical School as part of Singapore's strategy to become a leading centre for medical research and education.

2006: National Research Foundation set up.

2006: Experimental Therapeutics Centre set up.

2007: Singapore Institute of Clinical Sciences set up.

2007: Genentech, the world's first biotechnology company that pioneered the manufacture of synthetic human insulin, to develop a 1,000l E coli manufacturing plant for the production of Lucentis, a drug used for treating patients with wet age-related macular degeneration. It is Genentech's first plant outside of the US. Genentech is a fully owned subsidiary of Roche, one of the world's biggest pharmaceutical companies.

2008: Singapore Immunology Network set up.

2008: National Metrology Centre comes under A*STAR.

2009: Genentech Singapore bought up Lonza Biologics Singapore's large-scale cell culture manufacturing facility in Tuas Biomedical Park, merging it with Genentech Singapore's 2007 biologic manufacturing facility. In 2010, it was given the Facility of the Year award by the International Society for Pharmaceutical Engineering for its advanced pharmaceutical manufacturing features that can produce high quality drugs at lower costs.

2010: GlaxoSmithKline opened its first vaccine manufacturing plant in Singapore.

2010: Roche announced the setting up of a Translational Medicine hub with a budget of more than $130 million over three years with the aims of discovering and developing new and improved treatments in cancer and infectious diseases.

2011: Nanyang Technological University and DSO National Laboratories launched Singapore's first-ever micro satellite following nine years of R&D, one of the few countries in the region with its own satellite in orbit. It is named X-Sat.

2012: Inauguration of Singapore University of Technology and Design, in collaboration with MIT "to advance knowledge and nurture a new generation of technologically-grounded leaders — innovators equipped to create a better world through Design", officially opening its Sompah Road campus in 2015. It is to have "significant collaborations" with China's Zhejiang University.

2013: Singapore Centre for Nutritional Sciences, Metabolic Diseases and Human Development (SiNMeD) jointly established by A*STAR and NUS, a $148-million centre to focus on the union between scientific research, translational initiatives and clinical practice.

2013: Skin Research Institute of Singapore (SRIS) set up jointly by National Skin Centre, Institute of Medical Biology (IMB) and NTU. The centre evolved from the Skin Biology Cluster Platform and the Clinical Research Unit for Skin Allergy and Regeneration (CRUSAR) in IMB. The centre is to focus on skin biology.

2013: Materials Centre of Innovation (MCOI) with support from A*STAR and SPRING to help local SMEs better utilise technology innovation to increase their productivity and to boost material capabilities. The centre focuses on translating projects ready for further co-development with SMEs.

2014: Opening of Procter & Gamble's Innovation Centre in Biopolis, the largest private research facility in Singapore, to serve as P&G's international innovation hub for its hair, skin and home care, personal health and grooming products.

2015: Block 71 San Francisco set up to tap the interest of investors and US startups in the Southeast Asian market.

2015: Seagate opened a $100-million 500-man R&D facility.

2015: Takeda Pharmaceutical relocated its emerging markets HQ to Biopolis.

Chairmen, Executive Directors and Directors

NSTB's/A*STAR's Chairman/Deputy Chairman/CEO/MD

Designations	Names and Timeline
National Science and Technology Board (NSTB)/ Agency for Science Technology and Research (A*STAR) Chairman	1991–1993 : Lam Chuan Leong 1993–2001 : Teo Ming Kian 2001–2007 : Philip Yeo 2007– : Lim Chuan Poh
NSTB/A*STAR Deputy Chairman	1991–1999 : Prof Hang Chang Chieh 1998–1999 : Prof Su Guaning 1999–2000 : Philip Yeo 1999–2000 : Dr Finian Tan 2001–2003 : Prof Hang Chang Chieh 2004– : Prof Tan Chorh Chuan 2006–2007 : Lim Chuan Poh
NSTB/A*STAR Managing Director (MD)/ Executive Director (ED)	1990–1992 : Prof Chou Siaw Kiang (ED, NSTB) 1992 : Liew Mun Leong (ED, NSTB) Jan–Jul 1993 : Ms Leong Wai Leng (ED, NSTB) 1993–1996 : Vijay Mehta (ED, NSTB) 1996–1997 : Gong Wee Lik (ED, NSTB) 1998–2000 : Chong Lit Cheong (MD, NSTB) 2002–2006 : Boon Swan Foo (MD, A*STAR) 2007–2010 : Ms Yena Lim (MD, A*STAR) 2010–2012 : Prof Low Teck Seng (MD, A*STAR) 2012– : Prof Raj Thampuran (MD, A*STAR)
Biomedical Research Council (BMRC) Chairman	2001–2002 : Philip Yeo 2003–2007 : Dr Sydney Brenner 2007–2009 : Prof Sir David Lane 2009– : Prof Sir George Radda

(*Continued*)

(Continued)

Executive Director	2001–2002 : Prof Louis Lim 2003–2005 : A/Prof Kong Hwai Loong 2005–2006 : Prof Lam Kong Peng 2006–2008 : Dr Beh Swan Gin 2008–2011 : Prof Lee Eng Hin 2011– : Dr Benjamin Seet
Science and Engineering Research Council (SERC) Chairman	2001–2002 : Dr Sydney Brenner 2003–2005 : Philip Yeo 2005–2012 : Prof Charles F. Zukoski 2012– : Prof Sir John O'Reilly
Executive Director	2001–2002 : Prof Hang Chang Chieh 2003–2010 : Prof Chong Tow Chong Feb–Sep 2010 : Prof Low Teck Seng 2010–2013 : Dr Raj Thampuran 2013–2015 : Dr Tan Geok Leng
Exploit Technologies Pte Ltd Chairman	2002–2009 : Boon Swan Foo (Executive Chairman) 2010–2012 : Tan Gee Paw (Non-Executive Chairman) 2012–2013 : Prof Low Teck Seng (Non-Executive Chairman) 2013– : Dr Raj Thampuran (Non-Executive Chairman)
ETPL CEO	2009– : Philip Lim

BMRC RIs/RCs Executive Director

BMRC RIs/RCs	Executive Director (unless otherwise stated)
Bioinformatics Institute (BII)	2001–2003 : Dr Gunaretnam Rajagopal (Founding Director) 2003–2005 : Dr Santosh K. Mishra 2005–2007 : Dr Gunaretnam Rajagopal (ED) 2007– : Dr Frank Eisenhaber
Bioprocessing Technology Unit (BTU)/Bioprocessing Technology Centre (BTC)/ Bioprocessing Technology Institute (BTI)	1990–2011 : Prof Miranda Yap (Founding Director) 2012– : Prof Lam Kong Peng
Experimental Therapeutics Centre (ETC)	2006–2009 : Prof Sir David Lane (Founding CEO) 2009– : Prof Alex Matter, MD (CEO)
Singapore Genomics Program (SGP)/Genome Institute of Singapore (GIS)	2000–2001 : A/Prof Kong Hwai Loong 2001–2011 : Prof Edison Liu (Founding ED) 2012– : Prof Ng Huck Hui
Institute of Bioengineering and Nanotechnology (IBN)	2003– : Prof Jackie Ying (Founding ED)
Centre for Molecular Medicine (CMM)/Singapore Stem Cell Consortium (SSCC, became a division of IMB from 2011)/ Institute of Medical Biology (IMB)	2006–2007 : Prof E. Birgitte Lane (CMM) 2005–2007 : Prof Roger Pederson (Chairman, SSCC) 2007–2012 : Dr Alan Colman 2007– : Prof E. Birgitte Lane (ED, IMB)
Institute for Molecular and Cell Biology (IMCB)	1985–2001 : Prof Christopher Y H Tan (Founding Director) 2001–2004 : Prof Hong Wanjin (Acting Director) 2004–2007 : Prof Sir David Lane 2007–2010 : Prof Neal Copeland 2010–2011 : Prof Stephen Cohen (Acting ED) 2011– : Prof Hong Wanjin
Singapore Bioimaging Consortium (SBIC)	2005–2010 : Prof Sir George Radda (Founding Chairman) 2010–2012 : Prof Philip Kuchel (1st ED) 2014– : Prof Patrick J. Cozzone
Singapore Institute of Clinical Sciences (SICS)	2007–2014 : Prof Judith L. Swain (Founding ED) Apr–Dec 2014 : A/Prof Chong Yap Seng (Acting ED) 2015– : A/Prof Chong Yap Seng
Singapore Immunology Network (SigN)	2006–2014 : Prof Philippe Kourilsky (Founding Chairman) 2006–2008 : Prof Lam Kong Peng 2014– : Prof Laurent Renia

SERC RIs/RCs Executive Director

SERC RIs/RCs	Executive Director (unless otherwise stated)
Magnetics Technology Centre (MTC)/Data Storage Institute (DSI)	1992–1998 : Prof Low Teck Seng (Founding Director of MTC & DSI) 1998–2010 : Prof Chong Tow Chong 2010–2015 : Dr Pantelis Sophoclis Alexopoulos 2015– : Dr Tan Yong Tsong (Acting ED)
Process Analysis and Optimisation Enterprise (PAOE)/ Chemical and Process Engineering Centre (CPEC)/Institute of Chemical & Engineering Sciences (ICES)	1994–1998 : Prof Ching Chi Bun (Director) 1998–2000 : Prof Ching Chi Bun (Founding Director of CPEC of NUS) 2002– : Dr Keith Carpenter (Founding Director)
Centre for Computational Mechanics (CCM)/National Supercomputing Research Centre (NSRC)/Institute of High-Performance Computing (IHPC)	1994–1998 : Prof Lam Khin Yong (Founding Director of CCM of NUS) 1993–1998 : Mrs Tan Chee Kiow (NSRC) 1998–2003 : Prof Lam Khin Yong (Founding ED of IHPC) 2003–2005 : A/Prof Lee Heow Pueh (Acting ED) 2006–2010 : Dr Raj Thampuran 2010–2011 : Prof David Srolovitz 2012– : Prof Alfred Huan
Institute of Systems Science (ISS R&D Group)/Information Technology Institute (ITI)/Kent Ridge Digital Labs (KRDL)/ Centre for Signal Processing (CSP)/Laboratories for Information Technology (LIT)/Centre for Wireless Communications (CWC)/ Institute for Communications Research (ICR)/LIT + ICR = Infocomm Research Institute (I²R)	1985–1988 : Dr Ifay Chang (ISS, Founding Director) 1988–1998 : Dr Juzar Motiwalla (Director) 1986–1988 : Lim Swee Say (ITI, Founding Director) 1989–1991 : Mrs Chin Tahn Joo (Director) 1992–1994 : Dr Christopher Chia (Director) 1994–1995 : Dr Lim Joo Hong (Director) 1995–1998 : Dr Francis Yeoh (Director) 1998–2001 : Dr Juzar Motiwalla (KRDL, Founding CEO) 1995–1997 : Prof Er Meng Hwa (CSP, Founding Director) 1997–2002 : A/Prof Ser Wee 2002– : Dr Wong Lim Soon (LIT, Deputy Director) 1992–1993 : Prof Tjhung Tjeng Thiang (CWC, Founding Director) 1993–2002 : Prof Lye Kin Mun (Director) 2002 : Prof Lye Kin Mun (ICR, Deputy Director) 2002–2006 : Prof Lawrence Wong (I²R, ED) 2006–2010 : Managed by ExCo chaired by Prof Chong Tow Chong 2010–2012 : Prof Lye Kin Mun 2012–2013 : Dr Tan Geok Leng 2014– : Dr Lee Shiang Long

(Continued)

(Continued)

SERC RIs/RCs	Executive Director (unless otherwise stated)
Institute of Materials Research and Engineering (IMRE)	1996–1999 : Prof Shih Choon Fong (Founding Director) 1999–2000 : Dr William Chen (Acting Director) Jun–Aug 2000 : Dr Chou Sun Chee (Deputy Director) 2000–2003 : Prof Albert F. Yee (Director) 2003–2004 : ExCo 2004–2010 : Dr Lim Khiang-Wee
Institute of Microelectronics (IME)	1991–2000 : Dr Bill Chen (Founding Director) 2000–2004 : Dr Tan Khen Sang 2005– : Prof Dim-Lee Kwong
National Metrology Centre (NMC)	1987 : Teo Nam Kuan (Founding Head) 2008–2010 : Steven Tan (Director) 2010–2012 : Dr Lee Loke Chong 2012–2013 : Dr Thomas Liew (Acting ED) 2013– : Dr Thomas Liew
Grumman International and Nanyang Technological Institute on CAD-CAM (GINTIC)/GINTIC Institute of Computer Integrated Manufacturing (GICIM) Institute of Manufacturing Technology (IMT) GICIM + IMT = GINTIC Institute of Manufacturing Technology (GIMT)/ Singapore Institute of Manufacturing Technology (SIMTech)/GIMT + IMT = SIMTech	1985–1989 : James H. Boyd (Director of GINTIC) 1990–1993 : A/Prof Ho Nai Choon (Director of GINTIC Institute of CIM) 1990–1992 : A/Prof Yap Kian Tiong (IMT, Director) 1992–1993 : Dr Lee Loke Chong (Director) 1993–2002 : Dr Frans Carpay (1st Director of IMT/ GIMT/SIMTech) 2002–2005 : Dr Lim Khiang-Wee (1st ED) 2005– : Dr Lim Ser Yong

Honouring Scientific Talent

Honorary Singapore Citizen Award

As a way of honouring global talent that have made a marked contribution to Singapore in the fields of science, education and the economy, the Honorary Singapore Citizenship award was instituted in 2003. The Honorary Citizen Award is the highest form of state recognition that may be conferred on non-citizens. It ranks ahead of other National Day awards that may be conferred on non-citizens including the Public Service Star (Distinguished Friends of Singapore) and the Public Service Medal (Friends of Singapore). The Honorary Citizen Award reflects Singapore's welcome and recognition of foreign talent and their significant contributions to Singapore. It was first given out in 2003 and the first two recipients were:

2003: Dr Sydney Brenner and Pasquale Pistorio.
While 2002 Nobel Laureate Dr Brenner was a pioneer in the efforts to establish Singapore as a hub for biomedical sciences, Pistorio, an Italian and CEO of ST Microelectronics set up the first wafer fabrication plant in Singapore in 1984 and helped establish Singapore's semiconductor industry.

2004: Dr Sir Richard Brook Sykes
A pioneer in Singapore's Biomedical Sciences Initiative, Sir Richard helped to set up the Genome Institute of Singapore, and chairs the Biomedical Sciences International Advisory Council.

2006: Robert A. Brown played a key role in establishing a collaboration between National University of Singapore and Massachusetts Institute of Technology, which led to a number of highly-rated research and educational programmes in Singapore.

2011: Prof Edward W. Holmes
A pioneer in Singapore's translational and clinical sciences, Prof Holmes headed A*STAR's Translational and Clinical Sciences Group and chaired the National Medical Research Council of the Ministry of Health when translational biomedical sciences came onto the national R&D agenda.

2015: Prof Sir George Radda
The Hungarian chemist was recognised for playing a pivotal role in Singapore's biomedical sciences industry.

National Science and Technology Awards and Medals

The first-ever award for science and technology in Singapore was the Science Council's Gold Award for Applied Research presented in 1969 to Dr TG Ling, an industrial chemist who modernised pig and poultry farming with his scientifically balanced feeds. The Science Council introduced the National Science and Technology Award (NSTA) in 1987, the highest scientific and technology recognition of the time. In 1992, the National Science and Technology Medal was introduced and in 1993, National Science and Technology Award was split into two categories. In 2009, all three award categories under the NSTA umbrella were elevated into the current three categories under the President's Science and Technology Awards. The evolution of these top national honours became possible as Singapore's science and technology research landscape developed range and depth, became more nuanced and internationally significant. (For information on the awardees and their citations, go to the Agency for Science, Technology and Research website <a-star.edu.sg>.)

National Science and Technology Medal

1992 : Frank Cloutier
1993 : Prof Christopher YH Tan
1995 : Dr Herbert Eleuterio
        Dr Goh Hak Su
1997 : Sim Wong Hoo
1998 : Prof Chua Nam Hai
1999 : Prof Leo Tan Wee Hin
2000 : Prof Hang Chang Chieh

2001 : Dr Bill Chen

      Dr Frans Carpay

2002 : Prof Lui Pao Chuen

2003 : Prof Su Guaning

2004 : Kay Das

      Prof Low Teck Seng

2006 : Dr Sydney Brenner

2008 : Prof Tan Chorh Chuan

      Say Kwee Teck

President's Science and Technology Medal

2009 : Prof Miranda Yap

2010 : Prof Chong Tow Chong

2011 : Prof Soo Khee Chee

2012 : Prof Dim-Lee Kwong

2013 : Prof Freddy Boey

      Prof Barry Halliwell

2014 : Prof John Eu-Li Wong

2015 : Tan Gee Paw

National Science and Technology Award

1987 : Prof Wong Hock Boon

      Prof Lee Seng Lip

1988 : (Team award) Prof S S Ratnam, Dr Wong Peng Cheang, Dr Ng Soon Chye, Dr Chinnaiya Anandakumar, Dr Victor Goh, Dr Wong Yee Chee, Dr Arifeen Bongso, and Dr Clement Chan

1990 : Prof Lam Toong Jin

1991 : Dr Louis Chen Hsiao Yun

1992 : Prof Chan Soh Ha

National Science Award

1993 : Prof Ang How Ghee

1994 : Prof Huang Hsing Hua

1995 : Dr Catherine Pallen

Dr Peter Ng Kee Lin

1996 : (Team award) A/Prof Kang En Tang, A/Prof Neoh Koon Gee, and Prof Tan Kuang Lee;

(Team award) Dr Tay Sun Kuie, Dr Ang Peng Tiam, and Dr Hui Kam Man

1997 : Prof Goh Suat Hong

Prof Hew Choy Sin

1998 : Prof Gan Leong Ming

(Team award) Prof Lee Seng Luan, Dr Shen Zuowei, and A/Prof Wayne Lawton

1999 : Prof Hong Wanjin

2000 : A/Prof Sun Yeneng

2001 : (Team award) Prof William Chia, and Dr Yang Xiaohang

2002 : Prof Arif Bongso

2003 : (Team award) A/Prof Zhang Dong Hui, and Prof Lee Soo Ying

(Team award) Prof Harald Niederreiter, A/Prof Ling San, and A/Prof Xing Chaoping

2004 : Prof Alan Porter

(Team award) A/Prof Byrappa Venkatesh, Dr Samuel Aparicio, and Dr Sydney Brenner

2005 : A/Prof Li Baowen

A/Prof Zhang Lian Hui

2006 : (Team award) Dr Ruan Yijun, Dr Wei Chia Lin, Dr Patrick Ng Wei Pern, and Dr Ken Sung Wing Kin;

(Team award) Prof Oh Choo Hiap, Prof Berthold-Georg Englert, A/Prof Dagomir Kaszlikowski, and A/Prof Kwek Leong Chuan

2007 : Dr Ng Huck Hui

A/Prof Uttam Surana

(Team award) Prof AJ (Jon) Berrick, and A/Prof Wu Jie

2008 : Prof Mohan Balasu

(Team award) A/Prof Christian Kurtsiefer, A/Prof Valerio Scarani, Asst Prof Antia Lamas-Linares

President's Science Award

2009 : (Team award) A/Prof Aung Tin, Prof Roger Beuerman, and Prof Donald Tan

2010 : Prof Yoshiaki Ito

Prof Wong Tien Yin

2011 : (Team award) Dr Lim Bing, Dr Lawrence Stanton, Dr Paul Robson, and Dr Ng Huck Hui

(Team award) Prof Ooi Beng Chin, and Prof Tan Kian Lee

2012 : Prof Wang Yue

2013 : Prof Boris Luk'yanchuk

Prof Yu Hao

2014 : Prof Loh Kian Ping

2015 : (Team award) Prof Patrick Tan, Prof Teh Bin Tean, and Prof Steven Rozen

National Technology Award

1993 : (Team award) Dunstan Peiris, Dr Zhang Jian Guo, and Dr Tan Lye King;

(Team award) Dr Chew Tat Leong, Andrew Gill, Ang Chong Hoat, and Low Kee Haw

1994 : Chan Teong Lin

(Team award) Nixon Ng Ho Kwong, Pang Eng Poh, Dr Sun Jia, and Neui Yew Keng

1995 : (Team award) Neoh Chong Lim, and Chan Sang Kong

(Team award) Dr Ding Jeak Ling, and Dr Ho Bow

(Team award) Eddie Ko Beng Lee, and Dr Sun Zheng

1996 : (Team award) Daniel Lau Chin Hua, He Liang, Yuan Baosheng, and Lin Zhiwei

1997 : Quak Beng Wee and Team

1998 : Dr Lu Yong Feng

1999 : (Team award) Dr Chen Xiaoqi, Dr Gong Zhiming, Dr Huang Han, Dr Ge Shuzui, and Liau Soon Loong

2000 : A/Prof Lee Sing Kong

(Team award) Dr Shi Xu, Prof Tan Hong Siang, and A/Prof Tay Beng Kang

2001 : (Team award) Prof Lye Kin Mun, Prof Chow Shui-Nee, and Dr Jurianto Joe

2002 : (Team award) Dr Cheok Beng Teck, Prof Andrew Nee Yeh Ching, Dr Jiang Ridong, and Lim Tong Wah

(Team award) Dr Chang Soo Kong, Linda Wu Yongling, Leon Krings, and Sia Lee Heng

2003 : (Team award) Prof Tay Joo Hwa, A/Prof Tay Tiong-Lee Stephen, Asst Prof Show Kuan Yeow, and Asst Prof Liu Yu

2004 : (Team award) Dr Shi Luping, Dr Miao Xiangshui, and Tan Pik Kee

2005 : (Team award) A/Prof Zhang Mingsheng, Dr Liu Bo, and Leonard Gonzaga

2006 : (Team award) Dr Bi Chao, Dr Jiang Quan, and Dr Lin Song

2007 : (Team award) Huang Haibin, Dr Susanto Rahardja, Dr Yu Rongshan, and Dr Lin Xiao

(Team award) Jain Raj Kumar, Dr Sim Hak Keong, Dr Goh Chee Kiang, and Teo Tee Yong

2008 : (Team award) Dr Lo Guo-Chiang (Patrick), Dr Narayanan Balasubramaniam, Dr Navab Singh, and Dr Ajay Agarwal

(Team award) Dr John Yong MingShyan, Steven Tong Kin Kong, Dr Chua Beng Wah, and Ho Meng Kwong

President's Technology Award

2009 : (Team award) Prof Jacob CH Phang, Goh Szu Huat, Alfred CT Quah, and Chua Choon Meng

2010 : (Team award) Dr Patrick Lo Guo-Qiang, Dr Liow Tsung-Yang, Dr Ang Kah Wee, and Dr Yu Mingbin

2011 : Prof Lim Chwee Teck

2012 : (Team award) A/Prof Louis Phee, and Prof Lawrence Ho Khek Yu

2013 : (Team award) Prof Li Haizhou, Aw Ai Ti, Dr Su Jian, and Dr Ma Bin

2014 : (Team award) Prof Subbu Venkatraman, A/Prof Tina Wong, and Prof Freddy Boey

(Team award) Prof Wynne Hsu, Prof Wong Tien Yin and Prof Lee Mong Li

2015 : Prof Neal Tai-Shung Chung

Oral History Interviews

Oral History Interviews deposited at A*STAR's Office of Science Communications and ARchives (OSCAR) and the Oral History Centre of the National Archives of Singapore. Unless otherwise indicated, all interviews were conducted between 2014 and 2015 by Lee Geok Boi. Where indicated, some interviews were intended for this publication only. Researchers using oral history materials should note that such accounts give colour and human interest but must be cross-checked against contemporary written accounts such as annual reports and newspaper clippings.

Adams, Lee Ying

Brenner, Sydney

Carpay, Frans

Carpenter, Keith

Ching Chi Bun

Chong Tow Chong

* Chow, Edwin

* Halliwell, Barry

Hang Chang Chieh

Holmes, Edward

Hong Wanjin

Kong Hwai Loong

Lam Khin Yong

Lam Kong Peng

Lane, David

Lee Eng Hin

Lee Kum Tatt (interviewed by Daniel Chew in 1993)

Lim Chuan Poh

Lim Khiang-Wee

Lim, Louis

Lim Ser Yong

Ling, James

Liu, Edison

Lye Kin Mun

Low Teck Seng

* Png Cheong Boon

Radda, George

* Raj Thampuran

* Seet, Benjamin

Shih Choon Fong

Suresh Sachi

Sykes, Richard

Tan, Christopher, Yin Hwee

* Tan Kai Hoe

Tan Kok Kheng

Teo Ming Kian

Teo Nam Kuan

Wong, Lawrence, Wai Choong

** Yeo, Philip

Ying, Jackie

* Interviewed only for this publication.

** Interviewed for A*STAR's 20th anniversary publication in 2011.

Acronyms

This list of acronyms covers bodies, agencies, institutions, processes, scientific terms and other multi-word nouns that may have been mentioned in this publication in order to ease readers' reference to the plethora of acronyms that is today an essential part of literary life.

AAZ	—	Advanced Audio Zip
ACE	—	Action Community for Entrepreneurship
ACI	—	Institute of Asian Consumer Insights
AFEO	—	ASEAN Federation of Engineering Organisation
AGA	—	A*STAR Graduate Academy
AMD	—	Age-related Macular Degeneration
AMPL	—	Advanced Molecular Pathology Laboratory
APDS	—	Advanced Packaging Development Support
APIs	—	Active Pharmaceutical Ingredients
APS	—	Advanpack Solutions Pte Ltd
ARC	—	Applied Research Corporation
ASEAN	—	Association of Southeast Asian Nations
ASPA	—	Association of Singapore Patent Agents
A*STAR	—	Agency for Science, Technology and Research
ATREC	—	Advanced Technology Research Centre
BII	—	Bioinformatics Institute

BMRC	—	Biomedical Research Council
BMS	—	Biomedical Sciences
BTI	—	Bioprocessing Technology Institute
BTU	—	Bioprocessing Technology Unit
CAD/CAM	—	Computer-Aided-Design/Computer-Aided-Manufacturing
CBMM	—	The Center for Brains, Minds & Machines
CCM	—	Centre for Computational Mechanics
CITREP	—	Critical Infocomm Technology Resource Programme
CEO	—	Chief Executive Officer
CITREP	—	Critical Infocomm Technology Resource Programme
CMM	—	Centre for Molecular Medicine
CMOS	—	Complementary metal-oxide semiconductor
CNRC	—	Clinical Nutrition Research Centre
COI	—	Centres for Innovation
CORE	—	Centre for Offshore Research & Engineering
CPEC	—	Chemical and Process Engineering Centre
CQT	—	Centre for Quantum Technologies
CREATE	—	Campus for Research Excellence And Technological Enterprise
CRP	—	Competitive Research Programme
CSHL	—	Cold Spring Harbor Laboratory
CSI Singapore	—	Cancer Science Institute of Singapore
CSP	—	Centre for Signal Processing
CTO	—	Chief Technology Officer
CWC	—	Centre for Wireless Communications
DBS	—	Development Bank of Singapore
DPC	—	Development Project Committee

DR	—	Diabetic Retinopathy
DSI	—	Data Storage Institute
DSIR	—	Britain's Department of Science and Industrial Research
DSO	—	Defence Science Organisation
DSTA	—	Defence Science and Technology Agency
EC	—	Economic Committee
EDB	—	Economic Development Board
EEC	—	European Economic Community
EMERL	—	Electromagnetic Effects Research Laboratory
EOS	—	Earth Observatory of Singapore
EPC	—	Electronic Product Code
ERI@NTU	—	Energy Research Institute of NTU
ESF	—	European Science Foundation
ETC	—	Experimental Therapeutics Centre
ETPL	—	Exploit Technologies Pte Ltd
EURYI	—	European Young Investigator Awards
EWI	—	Environment and Water Industry Programme Office
FAR	—	Failure Analysis and Reliability
FBG	—	Fibre Bragg Grating
FDA	—	Food and Drug Administration
FSI	—	French-Singapore Institute
GDP	—	Gross Domestic Product
GE	—	General Electric
GERD	—	Gross Expenditure on Research & Development
GET-Up	—	Growing Enterprises through Technology Upgrade
GICIM	—	GINTIC Institute of Computer Integrated Manufacturing
GIMT	—	GINTIC Institute of Manufacturing Technology

GINTIC	—	Grumman International-Nanyang Technological Institute for CAD/CAM
GIS	—	Genome Institute of Singapore
GNP	—	Gross National Product
GSI	—	German-Singapore Institute
I&E	—	Innovation and Entrepreneurship
IAS	—	Institute of Advanced Studies
IBM	—	International Business Machines Corporation
IBN	—	Institute of Bioengineering and Nanotechnology
IC	—	Integrated Circuit
ICES	—	Institute of Chemical and Engineering Sciences
ICM		Infocomm Media
ICR	—	Institute for Wireless Communications
ICT	—	Information and Communications Technology
IDA	—	Infocomm Development Authority
IDM	—	Interactive and Digital Media
IDMI	—	Interactive and Digital Media Institute
IDS	—	Innovation Development Scheme
IE Singapore	—	International Enterprise Singapore
IEEE	—	Institute of Electrical and Electronics Engineering
IES	—	Institute of Engineers Singapore
IHPC	—	Institute of High Performance Computing
IMA	—	Institute of Molecular Agrobiology
IMB	—	Institute of Medical Biology
IMCB	—	Institute of Molecular and Cell Biology
IME	—	Institute of Microelectronics
IMM	—	Infocomm Media 2025 Masterplan

IMRE	—	Institute of Materials Research and Engineering
IMT	—	Institute of Manufacturing Technology
iN2015	—	Intelligent Nation 2015
IP	—	Intellectual Property
IPCF	—	Intellectual Property Competency Framework
IPI	—	Intellectual Property Intermediary
IPOS	—	Intellectual Property Office of Singapore
I²R	—	Institute for Infocomm Research
IRIS	—	Incentive for Research and Innovation Scheme
IRU	—	Industrial Research Unit
ISO	—	International Standards Organisation
ISS	—	Institute of Systems Science
IT	—	Information Technology
ITI	—	Information Technology Institute
JCO	—	Joint Council Office
JMI	—	Japan Machinery and Metals Inspection Institute
JSTI	—	Japan-Singapore Technical Institute
JTC	—	Jurong Town Corporation
KFELS	—	Keppel's Far East Levingston Shipping Ltd
KOM	—	Keppel Offshore & Marine
KPIs	—	Key Performance Indicators
KRDL	—	Kent Ridge Digital Labs
LETAS	—	Local Enterprise Technical Assistance Scheme
LIS	—	Light Industries Services
LIT	—	Laboratories for Information Technology
LLEs	—	Larger Local Enterprises
MAC	—	Microprocessor Applications Centre

MBI	—	Mechanobiolgy Institute
MD	—	Managing Director
MDA	—	Media Development Authority
MEMS	—	Microelectromechanical systems
MERS	—	Middle East Respiratory Syndrome
MINDEF	—	Ministry of Defence
MINT	—	Centre for Micromagnetics and Information Technologies
MIS	—	Moulded Interconnect System
MIST	—	Micro-Insemination Sperm Transfer
MIT	—	Massachusetts Institute of Technology
MNC	—	Multinational Corporation
MOE	—	Ministry of Education
MOF	—	Ministry of Finance
MOH	—	Ministry of Health
MOU	—	Memorandum of Understanding
MPA	—	Maritime and Port Authority
MPEG	—	Moving Picture Experts Group
MPT	—	Microelectronic Processing Technologies
MRI	—	Magnetic resonance imaging
MSA	—	Microelectronic System Applications
MTC	—	Magnetics Technology Centre
MTI	—	Ministry of Trade and Industry
mvtr	—	moisture vapour transmission rate
NCB	—	National Computer Board
NEWRI	—	Nanyang Environmental and Water Research Institute
NFIE	—	National Framework for Innovation and Enterprise
NGS	—	Next Generation Sequencing

NIC	—	Nikon Imaging Centre
NIST	—	National Institute of Standards and Technology
NMC	—	National Metrology Centre
NMRC	—	National Medical Research Council
NPB	—	National Productivity Board
NPI	—	*Nature* Publication Index
NRF	—	National Research Foundation
NS	—	National Service
NSRC	—	National Supercomputing Research Centre
NSTB	—	National Science and Technology Board
NTI	—	Nanyang Technological Institute (later to become NTU)
NTP	—	National Technology Plan
NTU	—	Nanyang Technological University
NUS	—	National University of Singapore
NWC	—	National Wages Council
oc	—	ovarian carcinoma
OEMs	—	Original equipment manufacturers
OSCAR	—	Office of Science Communications and ARchives
OTA	—	Operational Technology Assessment
OTD	—	Offshore Technology Development
OTR	—	Operational and Technology Roadmapping
P&G	—	Procter & Gamble
PAOE	—	Process Analysis and Optimisation Enterprise
PC	—	Personal Computer
PCBA	—	Printed Circuit Board Assembly
PCI	—	Preclinical Imaging
PCR	—	Polymerase Chain Reaction

PCT	—	Patent Cooperation Treaty
PDAS	—	Product Development Assistance Scheme
PIs	—	Principal Investigators
POC/POV	—	Proof-of-concept/Proof-of-value
POLARIS	—	Personalised OMIC Lattice for Advanced Research and Improving Stratification
PRIs	—	Public Research Institutes
PS	—	Permanent Secretary
PSB	—	Productivity and Standards Board
PUB	—	Public Utilities Board
R&D	—	Research and Development
RAD	—	Retinal Angiogenic Disease
RCEs	—	Research Centres of Excellence
RDAS	—	Research and Development Assistance Scheme
RDM	—	R&D Manager
REP	—	Renaissance Engineering Programme
REVIVE	—	Reverse Engineering Visual Intelligence for cognitiVe Enhancement
RF	—	Radio Frequency
RFID	—	Radio Frequency Identification
RIs	—	Research Institutes
RIE	—	Research, Innovation and Enterprise
RIEC	—	Research, Innovation and Enterprise Council
RISC	—	Research Incentive Scheme for Companies
RSEs	—	Research Scientists and Engineers
RTS	—	Research and Technical Support
SAF	—	Singapore Armed Forces

SARS	—	Severe Acute Respiratory Syndrome
S&E	—	Search and Examination
S&T	—	Science and Technology
SBIC	—	Singapore Bioimaging Consortium
SCELSE	—	Singapore Centre on Environmental Life Sciences Engineering
SDK	—	Software Development Kit
SEED	—	Startup Enterprise Development Scheme
SERC	—	Science and Engineering Research Council
SERI	—	Singapore Eye Research Institute
SET	—	Science, Engineering and Technology
SGH	—	Singapore General Hospital
SGP	—	Singapore Genomics Programme
SIA	—	Singapore Airlines
SICS	—	Singapore Institute of Clinical Sciences
SIgN	—	Singapore Immunology Network
SIMTech	—	Singapore Institute of Manufacturing Technology
SingTel	—	Singapore Telecommunications
SIPRAD	—	SERI-IMCB Programme In Retinal Angiogenic Diseases
SISIR	—	Singapore Institute of Standards and Industrial Research
SMEs	—	Small-Medium Enterprises
SMI	—	Singapore Maritime Institute
SNEC	—	Singapore National Eye Centre
SNP	—	Single Nucleotide Polymorphism
SONDRA	—	Supèlec ONERA NUS DSO Research Alliance
SPRING Singapore	—	Standards, Productivity, and Innovation for Growth

SRIS	—	Skin Research Institute of Singapore
SSCC	—	Singapore Stem Cell Consortium
SUTD	—	Singapore University of Technology and Design
T21	—	Technopreneurship21
TA	—	Technical Advisor
TAP	—	Technology Adoption Programme
TAS	—	Telecommunication Authority of Singapore
TCB	—	Thermo Compresion Bonding
TI	—	Texas Instruments
TP	—	Temasek Professorship
TRM	—	Technology Road Mapping
TSV	—	Through Silicon Via
T-Up	—	Technology for Enterprise Capability Upgrade
UHF	—	Ultra high frequency
UN	—	United Nation
UNESCO	—	The United Nations Educational, Scientific and Cultural Organisation
UV	—	Ultraviolet
VA	—	Value-add
VC	—	Venture capital
VIP	—	Visiting Investigatorship Programme
VLSI	—	Very Large Scale Integration
WHO	—	World Health Organisation
WIPO	—	World Intellectual Property Organisation

All dollar figures in this book refer to the Singapore dollar unless otherwise indicated.

Select Bibliography

A*STAR: 20 years of Science and Technology in Singapore. Published by Agency for Science, Technology and Research, [2011].

Agency for Science, Technology and Research
Annual report, 2001–

Chan Chin Bok (lead author). Heart Work: Stories of How EDB Steered the Singapore Economy from 1961 into the 21st Century. Singapore, Singapore Economic Development Board and EDB Society, 2002.

Chew, Melanie and Tan, Bernard
Creating the Technology Edge: DSO National Laboratories, Singapore, 1972–2002. Singapore, Epigram, 2002.

Chiang, Mickey
From Economic Debacle to Economic Miracle: The History and Development of Technical Education Education in Singapore. Singapore, Times Editions, 1998.

Creating Value: Singapore's Water R&D Journey. Singapore, PUB, 2013.

Fredberg, Tobias, Elmquist, Maria, and Ollila, Susanne
Managing Open Innovation: Present Findings and Future Directions. Chalmers University of Technology. VINNOVA Report VR 2008:02 [Stockholm], VINNOVA — Verket för Innovationssystem/Swedish Governmental Agency for Innovation Systems, 2008.

Friedberg, Errol C. Sydney Brenner: A Biography. Cold Spring Harbour, NY, Cold Spring Harbor Laboratory Press, 2010.

Lee, Edwin and Tan Tai Yong
Beyond Degrees: The Making of the National University of Singapore. Singapore, Singapore University Press, 1996.

Lim Swee Say and Chang, Ifay F.
National Computerization and Road Map to Information Technology Research and Development. Paper prepared by Lim Swee Say (Information Technology Institute) and Ifay F. Chang (Institute of Systems Science) for Institute of Electronic Engineers publication, 1987.

Nanyang Technological University
Annual report, 1992–

National University of Singapore
Annual report, 1981–

Ngiam, Tong Dow
A Mandarin and the Making of Public Policy: Reflections. Singapore, NUS Press, 2006.

150 Years of Education in Singapore. Singapore, TTC Publications Board, 1969.

Saravan Gopinathan
Towards a National System of Education in Singapore 1945–1973. Singapore, Oxford University Press, 1974.

Science & Technology Quarterly: A Quarterly Report on National and International Science and Technology Policies and Programmes. Published by Science Council of Singapore. Singapore, The Council, 1980–1987.

Singapore. Dept of Statistics.
Economic & Social Statistics Singapore 1960–1982. Singapore, 1983.

Singapore. Economic Development Board.
Annual report, 1962–

Singapore. Government Gazette. Acts Supplement
No. 1, 2 Jan 1991: The National Science and Technology Board Act 1990 (No. 24 of 1990)
Singapore, Government Printers, 1991.

Singapore. Ministry of Science and Technology. Economics and Statistics Unit
Directory of Scientific and Technical Research and Consultancy
Establishments in Singapore. Singapore, 1971.

Singapore. Ministry of Trade and Industry
The Singapore Economy: New Directions. Report of the Economic Committee. Singapore, 1986.

Singapore. Productivity and Standards Board
Annual report, 1997–

Singapore Institute of Standards and Industrial Research
Annual report, 1973/74–1995/96

Singapore. National Science and Technology Board
Annual report, 1992–2000.

Singapore. National Science and Technology Board
National Survey of R&D in Singapore 1978–
Triennial 1978–1987. Annual report, 1990–

Singapore. National Science and Technology Board
Annual report, 1991–2001.

Singapore. National Science and Technology Board
National Survey of R&D in Singapore 1990–
Annual report, ??

Singapore National Academy of Science
Trends in Science Education, Proceedings of Seminar. Eds: Kuok, MH, and Das, NP.
Singapore 1989.

Singapore Science Council
Annual report, 1968–1990.

Singapore Science Council
National Survey of R&D in Singapore 1978–1987. Triennial.

Singapore Science Council
Proposals for the Setting up of a Scientific Technical Information Centre in Singapore.
Report by Dr H Bauer. Singapore, 1969.

Singapore Science Council
Windows of the Future: A Status Report of the Singapore Science Park. Written by Guo
Yee Leng and Dr Mohand Narendran. Singapore, 1987.

Singapore Statutes
The Science Council of Singapore Act (Act 13 of 1967) and The Science Centre Act
(Act 33 of 1970). Rev. ed. Vol. VIII.

Internet, Non-book sources

Boyer, Herbert/Genentech
http://www.gene.com/media/company-information/chronology (Retrieved 19 June 2015)

Chesbrough, Henry. *Everything You Need to Know About Open Innovation* in *Forbes*,
March 2011
http://blogs.forbes.com/henrychesbrough/2011/03/21/everything-you-need-to-know-about-open-innovation/ (Retrieved May 2015)

Creative Technology
http://www.funize.com/company-histories/Creative-Technology-Ltd-Company-History.html (Retrieved 19 June 2015)

Duke-NUS Story
https://www.duke-nus.edu.sg/about/duke-nus-story

Economic Development Board
https://www.edb.gov.sg/content/edb/en/industries.html

Everbloom Mushrooms Pte Ltd
http://www.mycobiotech.com/milestones.html

National Human Genome Research Institute
http://www.genome.gov/10005139 (Retrieved 19 June 2015)

National Library Board SearchPlus
http://search.nlb.gov.sg (Retrieved 19 June 2015)

Singapore newspapers
http://eresources.nlb.gov.sg/newspapers/Default.aspx (Retrieved 19 July 2015)

National University of Singapore
http://www.nus.edu.sg/president/past_presidents/tohchinchye.php (Retrieved 14 Jan 2011)

OpenInnovation.eu
http://www.openinnovation.eu/topic/science/ (Retrieved 19 June 2015)

National Research Foundation
www.nrf.gov.sg/ (Retrieved 1 May 2015)

Procter & Gamble
http://news.pg.com/blog/company-strategy/SGIC#sthash.1UF8kfPF.dpuf

Singapore University of Technology and Design
http://www.sutd.edu.sg/speech_uic_sutdpresident.aspx

Sousa, Milton
Open Innovation Models and Knowledge Brokers. Case Study
www.openinnovation.eu/.../LowResIKMarch08Case%20Study.pdf (Retrieved 19 May 2011)

The Straits Times

Sydney Brenner
http://libfe.cshl.edu/archives/brenner/index.html
http://library.cshl.edu/personal-collections/sydney-brenner
National University of Singapore. Faculty of Science
Sixty Years of the Faculty of Science
National University of Singapore (1929–1989)
Singapore, 1990

University of Malaya
The Register of Graduates (1910–1961)
Kuala Lumpur, 1963

Yeo, Philip. Speech at OIST Symposium 2010 in YouTube
http://www.youtube.com/watch?v=rQeiVP5LRrc&NR=1 (Retrieved 25 May 2011)

Index

Made in the USA
Monee, IL
07 July 2026

56548178R00195